NEUROMETHODS

Series Editor
Wolfgang Walz
University of Saskatchewan
Saskatoon, SK, Canada

For further volumes:
http://www.springer.com/series/7657

Genetically Encoded Functional Indicators

Edited by

Jean-René Martin

Imagerie Cérébrale Fonctionnelle et Comportements, Neurobiologie & Développement (N&D), CNRS, UPR-3294, Gif-sur-Yvette, France

 Humana Press

Editor
Jean-René Martin
Imagerie Cérébrale Fonctionnelle et Comportements
Neurobiologie & Développement (N&D)
CNRS, UPR-3294
Gif-sur-Yvette, France

ISSN 0893-2336   ISSN 1940-6045 (electronic)
ISBN 978-1-62703-013-7   ISBN 978-1-62703-014-4 (eBook)
DOI 10.1007/978-1-62703-014-4
Springer New York Heidelberg Dordrecht London

Library of Congress Control Number: 2012943160

Printed on acid-free paper

Humana Press is a brand of Springer
Springer is part of Springer Science+Business Media (www.springer.com)

Preface to the Series

Under the guidance of its founders Alan Boulton and Glen Baker, the Neuromethods series by Humana Press has been very successful since the first volume appeared in 1985. In about 17 years, 37 volumes have been published. In 2006, Springer Science + Business Media made a renewed commitment to this series. The new program will focus on methods that are either unique to the nervous system and excitable cells or which need special consideration to be applied to the neurosciences. The program will strike a balance between recent and exciting developments like those concerning new animal models of disease, imaging, in vivo methods, and more established techniques. These include immunocytochemistry and electrophysiological technologies. New trainees in neurosciences still need a sound footing in these older methods in order to apply a critical approach to their results. The careful application of methods is probably the most important step in the process of scientific inquiry. In the past, new methodologies led the way in developing new disciplines in the biological and medical sciences. For example, Physiology emerged out of Anatomy in the nineteenth century by harnessing new methods based on the newly discovered phenomenon of electricity. Nowadays, the relationships between disciplines and methods are more complex. Methods are now widely shared between disciplines and research areas. New developments in electronic publishing also make it possible for scientists to download chapters or protocols selectively within a very short time of encountering them. This new approach has been taken into account in the design of individual volumes and chapters in this series.

Wolfgang Walz

Preface

Wishing to look into the brain and visualize the neuronal activity dates not from the past one or two decades (roughly the time that the modern brain imaging techniques have evolved) but from much longer ago. Indeed, as can be seen in some old carvings dating from the seventeenth century, looking into the brain to decipher our thoughts and how the brain integrates the sensory modalities and converts them to emergent functions that allow an organism to act and modify its environment was already an exciting challenge. The recently developed modern methods of neuronal imaging permit us to tackle this challenge in a more precise and powerful manner, especially by allowing the visualizing and recording of the neuronal activity of a single or group of neurons simultaneously.

In the last two decades, several different technologies have been developed with the aim of investigating and understanding how neurons or neural circuits operate collectively in the nervous system leading to coherent brain activity. Briefly, in the second half of the last century, electrophysiological techniques have been highly successful in investigating neuronal activity. Culminating in tools like the patch-clamp, it brought this analysis to the more precise and minute level of single-channel resolution. However, these approaches have been mainly dedicated to the resolution of single cell level activity and are rather limited in describing whole brain activity. In parallel, other electrophysiological techniques like electroencephalogram (EEG) have been developed to study the general activity of the brain, such as sleep waves of activity. Although these last approaches have been informative, they remained relatively poor when attempting to assign a specific role to a defined structure or group of neurons (structure/function relationship). In the last decade, several different optical imaging techniques, either based on various voltage or calcium dyes, or more recently on modified fluorescent or bioluminescent proteins (genetically encoded) that are sensitive to calcium, have been developed to study neuronal activity, and especially groups of neurons, with the goal of mapping and deciphering the neural code underlying major neurophysiological functions.

The aim of this book is to present the development of recent genetic techniques that allow for generating genetically encoded activity sensors to investigate neuronal activity. Each chapter describes a specific sensor and its utilization to study neuronal activity in a particular way. Though some of them look similar, notably in terms of monitoring calcium activity, each possesses its own characteristics of sensitivity and kinetics. Consequently, the development of particular devices used to visualize them is also specific. Chapters 1 and 2 describe a recently developed bioluminescence approach that allows for following in real time, in continuous and over a long-term period, the neuronal activity, either in a few precisely targeted neurons or in the whole brain. This still young but promising approach is just at its beginning and will surely reveal further new neurophysiological mechanisms within the brain. Chapter 3 describes the use of the various Genetically Encoded Calcium Indicators (GECIs) and, more particularly, the cameleon calcium sensor and its use to investigate the olfactory circuits. Chapter 4 aims to describe the synapto-pHluorin that monitors the synaptic vesicles release, and therefore the synaptic transmission, and its application to the olfactory system. The fifth chapter presents a sophisticated method to perform two-photon calcium imaging with GCaMP and, simultaneously, whole-cell patch clamp and loose-patch electrophysiological recordings in a walking fly. This double approach permits us to parallel what can be detected by calcium imaging compared to the electrophysiological recordings. Chapter 6 reports the use of cameleon to study some neurons implicated in courtship and sexual behavior. In Chap. 7, the authors report the use of GCaMP to study the physiological characteristics of the neurons of the Mushroom-Bodies (Drosophila brain structures implicated in the learning and memory). Chapter 8 precisely describes an engineering methodology to improve the kinetic and the sensitivity of calcium sensors, while, finally, Chap. 9 nicely describes a specific sensor that measures the variation of the cAMP (a critical second messenger signaling molecule) and its application to monitor the clock neuron network of the adult fly brain.

I hope that this book will convince readers of the importance of this still relatively young field of research and more particularly of the use of invertebrate model systems to perform brain imaging. I hope that it will fully testify to the current excitement about the development of these various techniques and what they could bring and reveal on the functioning of the nervous system. Hopefully, it will inspire students and researchers and will serve as a useful guide to those who want to start using these different brain imaging techniques and require a bit of guidance in how to choose the best technique to match the goal of their study.

I would like to thank the following colleagues for accepting to comment on portions of the manuscript: Jean-François Ferveur, Roland Strauss, François Rouyer, Marc Moreau, José Cancella, Dick Nässel, and Troy Zars. I also thank Lydie Collet for her help in some iconography works and the CNRS for its continuous support and especially for having supported my ambitious and risky project that was to develop the bioluminescence brain imaging technique applied to *Drosophila*. Finally, I would like to thank my wife, Beatrice, and my children, Valentine, David, and Félix, for their encouragement, support, and inspiration.

Orsay, France   *Jean-René Martin*

Contents

Contributors

JASPER AKERBOOM • *Howard Hughes Medical Institute, Janelia Farm Research Campus, Ashburn, VA, USA*

ROBERT A.A. CAMPBELL • *Cold Spring Harbor Laboratory, Cold Spring Harbor, NY, USA*

M. EUGENIA CHIAPPE • *Howard Hughes Medical Institute, Janelia Farm Research Campus, Ashburn, VA, USA*

RONALD L. DAVIS • *Department of Neuroscience, The Scripps Research Institute Florida, Jupiter, FL, USA*

ESTELLE DROBAC • *Neurobiologie des Processus Adaptatifs, Université Pierre et Marie Curie—Paris 6, CNRS UMR 7102, Paris, France*

ANDRÉ FIALA • *Molecular Neurobiology of Behaviour, Johann-Friedrich-Blumenbach-Institute, Georg-August- University of Goettingen, Goettingen, Germany*

EYAL GRUNTMAN • *Cold Spring Harbor Laboratory, Watson School of Biological Sciences, Cold Spring Harbor, NY, USA*

KYLE S. HONEGGER • *Cold Spring Harbor Laboratory, Watson School of Biological Sciences, Cold Spring Harbor, NY, USA*

VIVEK JAYARAMAN • *Howard Hughes Medical Institute, Janelia Farm Research Campus, Ashburn, VA, USA*

MASAYUKI KOGANEZAWA • *Division of Neurogenetics, Tohoku University Graduate School of Life Sciences, Sendai, Japan*

SOH KOHATSU • *Division of Neurogenetics, Tohoku University Graduate School of Life Sciences, Sendai, Japan*

BERTRAND LAMBOLEZ • *Neurobiologie des Processus Adaptatifs, Université Pierre et Marie Curie—Paris 6, CNRS UMR 7102, Paris, France*

KATHERINE R. LELITO • *Department of Molecular Cellular and Developmental Biology, University of Michigan, Ann Arbor, MI, USA*

LOREN L. LOOGER • *Howard Hughes Medical Institute, Janelia Farm Research Campus, Ashburn, VA, USA*

JEAN-RENÉ MARTIN • *Imagerie Cérébrale Fonctionnelle et Comportements, Neurobiologie & Développement (N&D), CNRS, UPR-3294, Gif-sur-Yvette, France*

JONATHAN S. MARVIN • *Howard Hughes Medical Institute, Janelia Farm Research Campus, Ashburn, VA, USA*

THOMAS RIEMENSPERGER • *Molecular Neurobiology of Behaviour, Johann-Friedrich-Blumenbach-Institute, Georg-August-University of Goettingen, Goettingen, Germany*

SILKE SACHSE • *Department of Evolutionary Neuroethology, Max Planck Institute for Chemical Ecology, Jena, Germany*

ORIE T. SHAFER • *Department of Molecular Cellular and Developmental Biology, University of Michigan, Ann Arbor, MI, USA*

ANTONIA STRUTZ • *Department of Evolutionary Neuroethology, Max Planck Institute for Chemical Ecology, Jena, Germany*

LIN TIAN • *Department of Biochemistry and Molecular Medicine, School of Medicine University of California, Davis, CA, USA*

LUDOVIC TRICOIRE • *Neurobiologie des Processus Adaptatifs, Université Pierre et Marie Curie—Paris 6, CNRS UMR 7102, Paris, France*

GLENN C. TURNER • *Cold Spring Harbor Laboratory, Watson School of Biological Sciences, Cold Spring Harbor, NY, USA*

THOMAS VÖLLER • *Neurobiology and Genetics, Julius-Maximilians-University of Wuerzburg, Wuerzburg, Germany*

DAISUKE YAMAMOTO • *Division of Neurogenetics, Tohoku University Graduate School of Life Sciences, Sendai, Japan*

DINGHUI YU • *Department of Molecular and Cellular Biology, Baylor College of Medicine, Houston, TX, USA*

Chapter 1

In Vivo Functional Brain Imaging Using a Genetically Encoded Ca²⁺-Sensitive Bioluminescence Reporter, GFP-Aequorin

Jean-René Martin

Abstract

In experimentally amenable organism models, several different physiological techniques have been developed to functionally record the neuronal activity, with the goal to map the neuronal circuitry and elucidate the neural code underlying major neurophysiological functions, such as olfaction, vision, learning and memory, sleep, locomotor activity, to name but a few. Apart from electrophysiological approaches, the main approach is optical imaging, principally based on the detection of changes in calcium concentration using fluorescent probes/sensors. The first generation of sensors was based on detecting calcium activity using fluorescent dye markers. The second generation, based on the development of genetically encoded fluorescent probes has allowed to precisely target the neurons of interest. However, because all of these approaches based on fluorescence require light excitation, deep structures of the brain still remain difficult to record. This means that the development of other alternative or complementary techniques is still worthwhile. Recently a novel bioluminescence approach has been developed, allowing to functionally map, in vivo, neuronal activity and circuitry. The aim of this chapter is to describe detailed protocols, from the genesis and the use of the GFP-aequorin probe, the setup, the recording and the analysis methods to perform in vivo functional brain imaging in *Drosophila*. Some original results that have been revealed by this new approach are also presented as well as discussion about the biological signification of the detected and recorded Ca²⁺-activity. Finally, advantages and constraints of using this approach compared to others are discussed.

Key words: Bioluminescence, Functional brain imaging, GFP-aequorin, Calcium, Genetic, *Drosophila*, Olfaction, Mushroom-Bodies

1. Background and Historical Overview

Generally, brain activity is considered as the result of both the integration of multiple sensory modalities, which gather information from the external world, and activity from the internal state (somatosensory system and homeostasis). However, brain activity does

Jean-René Martin (ed.), *Genetically Encoded Functional Indicators*, Neuromethods, vol. 72,
DOI 10.1007/978-1-62703-014-4_1, © Springer Science+Business Media, LLC 2012

not only integrate these multiple information but also generates "novel" or "intrinsic" activity to interact and even modify the surrounding external world. This last "intrinsic" activity is sometime considered as the "emergent function" of the brain. Thus, to understand such a complex system, it is necessary to not only study the neuronal activity at the cellular level resulting from sensory modalities, but also the emergent properties of their interaction. In such way, several different techniques have been developed with the goal of investigating how neurons and/or neural circuits operate, as an ensemble, in the nervous system, leading to coherent brain activity.

1.1. In Vivo Functional Brain Imaging

In the aim to study and decipher the neural code underlying major neurophysiological functions, in the last decade, several different optical imaging techniques, either based on voltage or calcium sensitive dyes, or on modified fluorescent proteins that are sensitive to calcium, have been developed to record neuronal activity (1–4). These various techniques have been applied to different organism models, such as the nematode *Caenorhabditis elegans* (5), the fly *Drosophila melanogaster* (6–15), the zebrafish (16), and more recently in the mouse (17–20). However, the fluorescent light emission requires light excitation. Consequently, these experimental techniques can be hampered by autofluorescence, photobleaching, and phototoxicity (for review see (21)). Photobleaching and phototoxicity limit the duration of recording, while the autofluorescence, which reduces the signal to noise ratio, impedes the recording of deep structures and consequently of the whole brain. Additionally, the animal needs to be tightly fixed, which excludes some analysis, like recording from a freely moving animal. Hence, because of these side effects, the in vivo neuronal imaging of the brain, particularly the structures that are located deep in the brain, remains difficult to access.

1.2. Bioluminescence

Bioluminescence is light produced by an enzymatic reaction, for which the light excitation is not required. Although the light signal produced by bioluminescence is generally weak (compared to fluorescence), the signal-to-noise ratio is very good, even excellent, because the background is more or less nil. Several years ago, Shimomura and colleagues (22–24) have identified and isolated the GFP protein from the jelly fish. In the natural jelly fish, the GFP forms a molecular complex with the aequorin, a Ca^{2+}-sensor protein. Aequorin possesses three EF-hand structures characteristic of Ca^{2+}-binding sites. The binding of Ca^{2+} to aequorin induces a conformational change resulting in the oxidation of its cofactor, the coelenterazine in coelenteramide, via an intramolecular reaction, while consequently blue light is emitted (λ max: 470 nm) (22). The binding of Ca^{2+} to aequorin produces chemiluminescence (equivalent to the term bioluminescence), which allows

(Ca^{2+}) to be monitored in living organisms (24). This "isolated" and simple form of aequorin alone (without its GFP partner) has already been used for monitoring Ca^{2+}-waves during fertilization or gastrulation in organisms like Xenopus (25) and zebrafish (26). However, because of the weak light emission of the aequorin, it was difficult or even impossible to use it to undertake in vivo whole brain imaging to map the temporal dynamics of Ca^{2+}-signaling in entire neural structures, since such studies require a fast (in the range of millisecond to about 1 or 2 s) temporal resolution. Consequently, studies have been carried out only on cell populations or on dissected tissue preparations, by measuring its light emission in vitro with a luminometer, which allows long-time integration of the light (in the range of minutes) (27–31).

In the goal to increase the efficiency of the aequorin, and notably its light emission, few years ago, P. Brûlet and colleagues have designed a chimeric protein based on a fusion between the Green Fluorescent Protein (GFP) and the aequorin (32), similarly to the native complex "GFP-aequorin" previously identified several years ago in jelly fish (22–24). GFP-aequorin (GA) is a non toxic genetically encoded Ca^{2+}-sensitive bioluminescent reporter with improved light emission as a result of chemiluminescence resonance energy transfer (CRET or BRET phenomenon). Compared to aequorin, the light-emitting activity is increased from about 19 to 65 times (Table 1 in ref. (32)). Two main reasons have been raised to explain this increase: (1) the excited state energy of Ca^{2+}-bound aequorin is nonradiatively transferred to the GFP moiety (CRET), which red-shifts the wavelength of aequorin's light emission, (2) aequorin is more stable when expressed as a hybrid protein (32). Similarly to aequorin, the bioluminescence reaction of GA occurs within milliseconds after Ca^{2+}-binding and allows monitoring changes in Ca^{2+}-concentrations over a wide dynamic range from 0.1 μM to 1 mM (33, 34). GA is also relatively insensitive to pH of physiological range (35). Moreover, since this new chimeric protein has a GFP-moiety, it can be previsualized using a fluorescent light. Then this permits to precisely refine the focus on the desired structures to be recorded, as well as to take a reference image (that can be used further to overlay with the bioluminescent image). In addition, since it is genetically encoded, the expression of the GFP-aequorin probe can be precisely targeted to different structures or tissues of interest. Therefore, this new Ca^{2+}-reporter is sufficiently powerful to allow functional mapping of neuronal circuits, noninvasively (34, 36–40). In parallel, technological improvements principally based on photon detectors have been developed permitting to detect and record such relatively weak light emission.

1.3. Several Genetic Tools Provided by Drosophila

Drosophila is an excellent genetically tractable model system. Its genome is entirely sequenced (41) and its number of genes is estimated at about 13,600. Among several genetic tools that have been developed, the most remarkable is the binary P(GAL4) system (42), and its multiple variants (for a schematic drawing, see (43)). Briefly, the P(GAL4) system permits the specific targeted expression of a foreign gene in a precise and given tissue, groups of cells or neurons of interest. Then, based on the driver line (enhancer-trap or alternatively under the control of a specific DNA genomic regulatory sequence) to drive the expression of the GAL4 transcription factor, which consequently activates the transcription of a reporter gene placed downstream of its UAS activating sequence, any gene of interest can be expressed. In such a way, any genetically engineered functional reporter gene, such as cameleon, GCaMP (successive different versions), or more recently the GFP-aequorin (P(UAS-GFP-aequorin)), can be specifically expressed in a given specific group of neurons. More recently, other alternative binary expression system, as LexA (44) and more recently Q-system (45) have also been developed allowing expressing reporter genes. The main advantage provided by these new alternative systems is that in combination with the P(GAL4) system (simultaneously expressed in the same fly), a different group of cells (neurons or glial cells) could be simultaneously targeted. Thus, while we could record the neuronal activity in a given group of neurons, a distinct group of neurons (that potentially could interact with those recorded) could be either blocked or excited independently of the recorded neurons. To achieve such neuronal manipulation, several genetic tools have also been developed to specifically either block the synaptic transmission, e.g., *tetanus toxin* (46, 47) and/or *shibire* (48), or activate neurons, e.g., Trp1A (49), TrpM8 (50) (for a more complete list of different genetic tools available to manipulate the neurons activity, see the reviews (51, 52)). Finally, as aforementioned, all the *Drosophila* genes have already been "bioinformatically" identified. Consequently, two different laboratories have generated interferential RNA (RNAi) directed against each gene (covering the putative 13,600 genes of the entire genome). These RNAi have been placed under the control of the UAS sequence (P(UAS-RNAi)), and transgenic flies have been generated. Thus, in combination with the P(GAL4) system, any given gene can be specifically knocked-down (mutated) in specific groups of cells or neurons. These two RNAi libraries (NIG, Japan: (http://www.nig.ac.jp/english/index.html) and VDRC, Vienna, Austria (http://stockcenter.vdrc.at/control/main)) are available to the scientific community. Therefore, under the control of the P(GAL4) system (or eventually another binary system), a specific RNAi could be simultaneously expressed with the Ca^{2+}-sensor gene (e.g., GFP-aequorin) within the same neurons, allowing to knock-down a given gene in the same neuron than the one we are directly recording.

2. Equipment, Materials, and Setup

2.1. Materials

2.1.1. GFP-Aequorin Construct and Transgenic Flies

The insert of the chimeric gene GFP-aequorin (GA) of pG5A, which contains five copies of the linker between the GFP and the aequorin (32) was cut out with EcoRI and XhoI and inserted into the pUAST vector (42). Transgenic flies, through germ-line transformations of *white* (w^{1118}) flies were generated using standard techniques. Three independent transformation lines have been obtained, one inserted on the second chromosome (GA1), and two on the third chromosome (GA2 and GA3). Using the binary P(GAL4) expression system, the resulting transformants can be crossed with any specific GAL4 driver line (e.g., OK107, a MB-specific driver (53) to examine the expression pattern). The three transformation lines were shown to have similar fluorescence (due to the GFP-moiety of the GA). In addition, the three lines have been tested for their bioluminescence and have been found to give similar results. For practical reasons, the line UAS-GA2 is commonly uses for studies (38). In *Drosophila*, we also dispose of several P(GAL4) lines allowing to express any gene of interests in almost any structures of the brain, as the MBs, CC, antennal lobes, etc. In addition, as aforementioned, alternative binary expression systems as LexA system and Q-system are now available. We have already generated the LexA-operator:GFP-aequorin construct, and express it in the olfactory receptor neurons (ORNs). Preliminary results obtained in our laboratory indicate that this system (Or83b-LexA/LexA-operator:GFP-aequorin) is sufficiently sensitive to record odor-induced Ca²⁺-activity in the axon terminal of the ORNs.

2.1.2. Fly's Ringers

Several different (but related) fly's ringers have been developed and can be found in the literature (54, 55). For the bioluminescent recording, we used the fly's ringers developed by Fiala et al. (6): 130 mM NaCl, 5 mM KCl, 2 mM $MgCl_2$, 2 mM $CaCl_2$, 36 mM Sucrose, 5 mM Hepes-NaOH, pH adjusted precisely at 7.3. The fly's ringer is preferably made fresh every day, but it can be stored at 4°C for one or 2 days.

2.1.3. Coelenterazine

Coelenterazine (Fig. 1) the aequorin cofactor, forms a molecular complex with the aequorin, which permits to the aequorin to adopt its tertiary configuration (22). When the (Ca²⁺) increases in the vicinity of this molecular complex, the aequorin catalyzes the reaction of the coelenterazine into coelenteramide and emits a photon in the blue wavelength. However, because of the weak efficiency of this molecular reaction (generally only few photons are emitted by one molecule), this system (the aequorin alone) cannot be used as a reporter gene to perform functional, in vivo brain (neuronal) imaging. Indeed, although it has been used as an in vitro reporter gene (quantified within a luminometer), because of its weak efficiency, it was too weak to be used in real time (in a range of few seconds) to perform in vivo brain imaging.

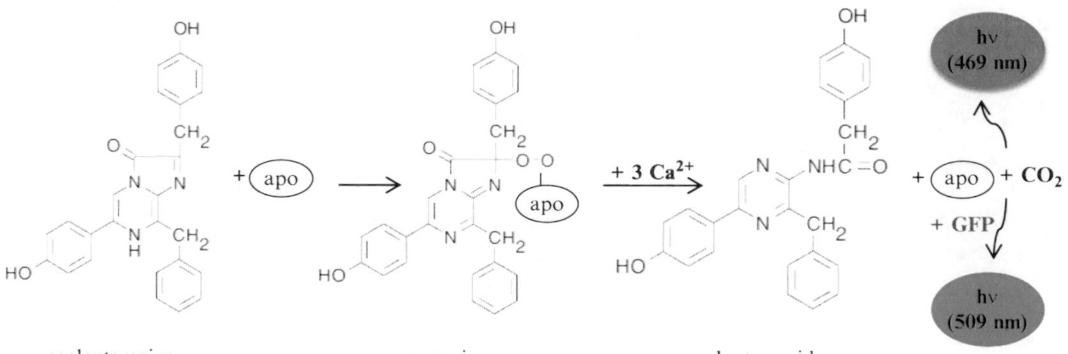

Fig. 1. Schematic drawing of the bioluminescent reaction between the coelenterazine and the GFP-aequorin. The cofactor coelenterazine binds to the apo-aequorin, which adopts its tertiary conformation. In presence of Ca^{2+} (upon binding of three atoms of Ca^{2+}) the aequorin catalyzes the coelenterazine into coelenteramide, and releases carbon dioxide (CO_2) and blue light. However, in presence of the GFP, a CRET phenomenon occurs and the light emission is shifted in the green wavelength (22, 32).

Several types of chemically modified analogues of coelenterazine have been generated (see Table 1 in ref. (23)), and are commercially available (Note 1). In the first series of experiments (38–40), we have used the native coelenterazine (Interchim, France) at final concentration of 5 μM. More recently, we use the benzyl coelenterazine (Cat# 303 NF-CTZ-FB, Prolume, USA), at a stock solution of 5 mM diluted in ethanol 100%, stored at −80 °C, and diluted just prior experiment at 5 μM. The coelenterazine is applied in the Ringer's solution and the tissue (brain) incubated for about 2 h (Note 1). A disadvantage is that in *Drosophila*, since the coelenterazine does not easily crosses the neural sheath (the equivalent of the brain blood barrier in mammals), this tight tissue might be either removed or made permeable. It can be delicately dissected by hand, or alternatively, as performed by the electrophysiologists (56), removed enzymatically, by using papain (papain incubation for 10 min at 50 U/ml).

2.2. Setup (Combined Fluorescence/ Bioluminescence Detection)

Since the low level of light emitted by the bioluminescent Ca^{2+}-sensor GFP-aequorin remains the major limiting factor (constraint) in the light detection, it is extremely important and crucial that the wide-field microscopy system be housed in a light-tight dark box (to avoid any external parasite light) (Fig. 2). As for the majority of

Fig. 2. (continued) Inc., MA, USA). A microscope is equipped with an EMCCD camera (Andor, Ixon), which can be driven in two modes: standard image acquisition and photon detector mode. The standard image acquisition mode is used to take the brightfield and the fluorescent image of GFP-aequorin expression (GFP moiety) in the brain to be recorded. This image will serve as a reference image. With the second mode (photon detector), each photon emitted are detected and their *x*, *y* coordinates as well as their time (*t*) are recorded. A fully automated multi-channel perfusion system allows pharmacological manipulations (similar to a perfusion bath). The multi-channel odor delivery system is also visible. The overall system is surrounded by a *tight light–dark box*.

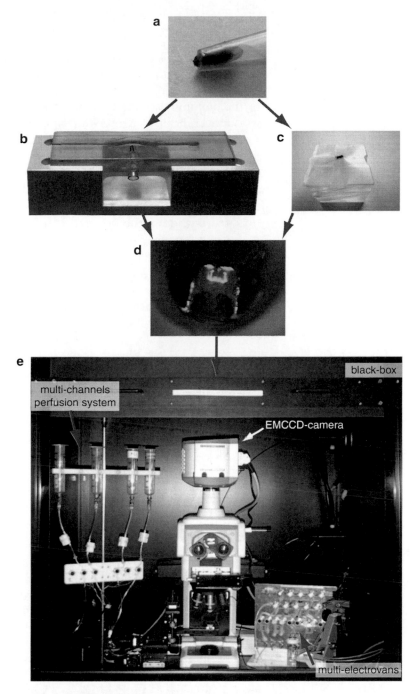

Fig. 2. Fly preparation, setup and accessories. (**a**) The fly is inserted on a *truncated blue tip* (1 ml) and fixed with glue. (**b**) For the snorkeling method used to perform pharmacological studies, the whole preparation (fly within the *blue tip*) is inserted in the hole of the chamber, which is filled with *Drosophila's* Ringer (up to a level to cover the head). (**c**) For the olfaction experiments, the fly within the blue tip is disposed on an acrylic block holder, and covered (tightly wrapped) with parafilm. (**d**) The head-cuticule of the fly is then opened, the fat body removed as well as the trachea. At this stage, the fly's brain in incubated for 15 min with papain (50 U/ml during 15 min) to permeabilize the neuroepithelium. Thereafter, the brain is incubated for 2 h with the coelenterazine (in a *humid dark box*, to avoid desiccation and light). (**e**) Setup used to detect the bioluminescence light emitted by the Ca^{2+}-activated GFP-aequorin. It consists of a dual fluorescence/bioluminescence wide-field microscope system custom built by Sciences Wares (Sciences Wares

in vivo imaging, it is recommended that the microscope-system is installed on an anti-vibratory table, to avoid any movement and distortion of images during the acquisition (particularly relevant for the bioluminescence system, since it allows long-term imaging period over hours). Moreover, since the microscope/system is enclosed in the dark box, it needs to be fully automated (the filter slider, the objectives and the three dimensions (x, y, z) stage), allowing to be fully driven outside the dark-box, by the appropriated software. It also requires that both standard and HBO arc lamps and their mechanical shutters control illumination are mounted outside of the box and connected via fibers optic guides to the microscope. Since up to now, such custom made microscopes are not directly available by standard custom companies, then a custom microscope has to be hand-modified (for example, in our laboratory, performed by ScienceWares Inc., MA, USA). To collect light, two different systems have been developed, as outlined in the next two sections.

2.2.1. Photon Detector

The first system was based on the use of an Photon Detector (for example, IPD 3, Photek Ltd., East Sussex, UK), which has been used at the Pasteur Institute, Paris, where the first studies have been performed (34, 36, 38, 57) (for a schematic drawing, see Fig. S1 in ref. (38)). The Photon Detector is connected to the C-port of the microscope, which assigns an x, y-coordinate and time point for each detected photon (58). The system is fully controlled by the data acquisition software, which also converts single photon events into an image that can be superimposed with brightfield or fluorescence images made by a connected standard CCD camera (for example, Coolsnap HQ, Roper Scientific) to a parallel second port. A software controlled automated-motorized mirror allows switching between the CCD camera and the Photon Detector. One of the main advantages is that the IPD produces almost no noise, or very low background counts (<1 photon/s in a 256×256 pixel region) (see Note 2 for detailed comparison with the EMCCD camera). Observations were made using a 20× objective with a highest numerical aperture (e.g., 0.5, Carl Zeiss, Germany). Resolution of the system was 256×256, with each pixel being equal to $12,875\ \mu m^2$. Theoretically, the data acquisition of a Photon detector could be made at a very high temporal resolution (in a range of millisecond up to the μs). However, up to now, mainly because for software and hardware interface constraints, the acquisitions have been made, at the best, at 50 ms time window (with the Pasteur Institute setup) (34, 38).

2.2.2. EMCCD Camera

An alternative to the Photon Detector system is based on the use of a single EMCCD camera which can be driven in two modes: image acquisition (for fluorescence) and photon detector mode (for bioluminescence) (similarly to the Photon detector) (Fig. 2). The single EMCCD camera can process sequentially both the acquisition of the image (fluorescence and brightfield) and the bioluminescent acquisition. For example (in our laboratory), the electron multiplier CCD camera (EMCCD, Andor, iXon, cooled to $-80°C$) is fitted onto the microscope (Nikon, Eclipse-E800). The setup is housed inside a tight dark box (Sciences Wares, Inc., USA). Using a 20× dry-objective lens (N.A., 0.75, Plan Apochromat, Nikon) (Note 3) the field of view is 400×400 μm (512×512 pixels). To improve signal to noise ratio data, a 2×2 binning can be used (1 pixel = 1.56×1.56 μm). The data can be acquired at a temporal resolution of any time resolution, from roughly 5 s up to 50 ms (also here, up to now, the time resolution is limited due to the software/hardware interface). However, in contrary to the Photon Detector, increasing the acquisition time (decreasing the temporal resolution) decreases the background level, because with the EMCCD camera, a steady "white reading noise" is generated at each acquisition (reading and resetting the camera) (see Note 2 for advantages/disadvantages). However, the acquisition and the storage of the data are similar to the Photon Detector: each detected photon is assigned x, y-coordinates and a time point (t). The main advantage of this acquisition/storage of the data, compared to the different fluorescent setups, which acquire and store data as images (generally in a tiff format), is that this requires very little space (storage memory) compared to tiff images. Moreover, since the acquired files are relatively small, they are easy to manipulate and fast to analyze (see section below).

2.3. Other Accessories

2.3.1. Electrovan (Odor Delivery Apparatus)

To perform the olfactory studies, we have developed a laboratory custom made odor delivery apparatus which consists of five identical channels (Fig. 2). One is devoted to control air (without odor), while the four others are used to deliver odors. Each channel comprises a 50 ml bottle with on either side a solenoid activated pinch-valve (Sirai S-104) isolating those not in use. An air pump continuously flows air at a rate of 500 ml/min, first through a moistening bottle (1 l) and after through the control channel except when a logic command issued by the imaging software switched the flow for the predetermined odor-duration time (of any given duration) (e.g., 1, 3 or 5 s or even as long as wished) to one of the test (odor) channels. The test flask contains a given volume of an odor (at the desired concentration) (e.g., 50 μl of undiluted pure odor) disposed on a piece of filter paper. Finally the air stream was delivered to the fly's antennae through a small tube placed a few millimeters away. Response of individual flies to different odors (e.g., spearmint, citronella, octanol) was recorded (39, 40). All connecting tubes are made of silicone, to avoid any interaction with odors.

2.3.2. Electric Stimulation

For electric stimulation, electric pulses were delivered by an isolated stimulus generator (Digitimer Mark II) triggered by the imaging software, through a glass micropipette previously filled with fly's Ringer. Different approaches can be used to deliver electric stimulation. In our study (39, 40), electric stimulation has been applied directly on the antennae. One antenna was sucked in by aspiration after sectioning the aristae. In that case, since the fly cuticle is thick and therefore acts as a very good electric insolent, it requires a relatively strong electric pulse (100 V, during 1 s). Alternatively, some studies, particularly the ones performed by the electrophysiologists (59, 60), are based on the opening of the antennae and anchors, with a hooked tungsten electrode, the antennal nerve to deliver directly the electric shock. In such a way, a much milder electric shock can be delivered. Finally, for the aversive associative olfactory learning and memory experiments, electric stimulation can be delivered through a tungsten microelectrode on the abdomen, or legs, or elsewhere in the fly, to associate it with an odor.

2.3.3. Locomotor Activity Recording Ball

To simultaneously record the in vivo neuronal activity within the brain and the locomotor activity of the fly, in the aim to study and map the brain structures that are involved in the control of locomotor activity, a small device, based on freely rotator ball sustained by a gentle airstream has been developed. The fly is installed and adjusted just above a tiny ball (generally made from styrofoam), which is maintained by a gentle airstream. The movement of the ball, which rotates when the fly is walking, is quantified by an infra-red light beam (based on a slightly modified infrared mouse-computer system) (N.B., the infra-red light emitting has been carefully selected in very narrow emission wavelength). In parallel, a selected infrared filter (with a very narrow bandwidth) has been added on the microscope, to completely cut-out all the infrared light, and avoid any light contamination. In complement, a software which records and quantifies the x and y movement of the mouse, has been developed. This system is currently functional in our laboratory (unpublished results) (for two different variants of the quantification of the walking activity, see also the two accompanying chapters (Chaps. 5 and 6)).

3. Procedures (or Methods)

3.1. Fly Preparation and Dissection (Brain Preparation)

Female flies at 4 days-old are generally used, as standard for our in vivo brain imaging experiments. We generally use females because they are slightly bigger than males (but in principle, there is no reason not to use males). We use 4-days-old because they are sexually mature and still young (healthy and strong) flies. However, it has no indication to not using older flies, the only importance is that flies of the same age might be compared between them (although up to now, it has no direct study demonstrating a difference between young versus older flies). The fly is briefly cold anesthetized on ice, inserted in a truncated 1 ml commercial (blue) pipette tip until the head protruded and finally fixed in place with dental glue (Protemp III, ESPE) (Fig. 2). The assembly was then placed in a custom-made acrylic block and secured with parafilm™. A drop of Ringer's solution (38) is deposited on the head, and a tiny window in the head capsule is cut out to expose the brain (e.g., forceps No. 5, knifes 22.5°, ref. 72-2201, FST Fine Scientific Tool, Germany). For the olfaction dedicated experiments, care is taken to not damage the antennae. Exposed brains are incubated in fly Ringer's solution containing 5 μM coelenterazine for minimum 2 h, before experiments (Note 4).

3.2. Imaging Preparation

Different types of experiments are currently performed in our laboratory. The imaging preparation depends of the purpose of the experiments. Briefly, two main types of preparation have been developed: (1) the snorkeling methods, more adapted and suitable for the pharmacological approach, and (2) a semibehavioral approach, allowing use of natural stimuli like odor presentation, or locomotor activity studies, or even for long-term imaging period.

3.2.1. Snorkeling Method (Pharmacology)

The fly is fixed as described above. The blue tip is inserted into the hole of a custom made chamber, and sealed (with silicone). The chamber is filled with fly's ringer, and the head capsule is opened, to expose the desired brain structures (Fig. 2). The preparation is then incubated with coelenterazine in ringer, in a humidified closed dark box, for 2 h. In this position, the fly can breathe by the tracheal system (located on each segment of the body) and can live for up to 24 h. The preparation is installed under the microscope, and using a water immersion objective (generally a 20×), the brain structures are visualized and imaged at desired time resolution (currently at 250 ms). The drugs (agonists, antagonists, etc.,) are delivered through a multi-channel perfusion system (e.g., Warner, USA). Alternatively, drugs could be applied by microinjection (e.g., nanoject or picospritzer) directly onto the brain structures, using a pulled-glass microelectrode, positioned with a motorized micromanipulator.

3.2.2. Semibehavioral Method (Upright System)

In the snorkeling method, since the fly has the opened head immersed in the Ringer, all studies based on the use of natural stimuli, like odors, are then excluded. To perform olfactory studies, the fly needs to have intact and exposed antennae. For this purpose, the fly is fixed as describe above. However, in this preparation, the blue tip containing the fly is installed on an acrylic block (Fig. 2). The head is wrapped with parafilm™ to seal the upper part of the head capsule. A drop of ringer is deposed and the head capsule dissected (opened) by hand. Coelenterazine is applied in the drop of ringer, and incubated for 2 h in a humidified dark-box. A glass coverslip is applied on the head, and the whole preparation installed under the microscope and visualized with a dry 20× objective. Generally, since dry objective possesses a better numerical aperture than the water-immersion, when possible, it is preferable to use dry objective. For the long-term recording period, as overnight (e.g., for the recording of the spontaneous activity in the Mushroom-Bodies: see below) or even longer (Note 5) we use a similar preparation, except that the fly is fed, by providing sugar-water.

3.3. In Vivo Brain Imaging: Data Acquisition and Analysis

Data acquisition can be performed at different chosen temporal resolution, from few seconds (which is relatively long) down to about 10 ms (Note 6). Imaging data and analyzing data software using the Photon Viewer (1.0) software, written in LabView 7.1 (National Instruments), has been developed under our requests and needs, by Sciences Wares (Sciences Wares, Inc., USA). For the acquired data, only the pertinent information is stored. Indeed, for each emitted photon, the x, y coordinates and the time of emission is stored (x, y, t). This has the main advantage that it requires only little memory/space in the computer (compared to the acquisition of successive images, taken by a standard camera, as for instance in a tiff format). Then, for example, for a continuous acquisition at 1 s, during 24 h, the size of a file is about only 50 MB.

For the analysis method, we use a second software that allows us to visualize and analyze the acquired data. Several parameters can be settled and adjusted, as drawing different ROIs of desired size, the accumulation time of x, y, z data to visualize the image, and the speed of analysis progress. The analysis is generally performed in two steps. First, by adjusting the accumulation time in a relatively large time window (for example, 30 s), allowing visualizing adequately the Ca^{2+}-signal (in such a condition, even a weak signal can be observed). Then, it also allows one to precisely draw the desired ROI on the structures that need to be analyzed. Afterwards, the proper analysis is done (at the same temporal resolution as the acquisition), and the data stored in a text format file, that can be used and analyzed further using any other softwares, like Excel™ (Microsoft), Statistica™, or others. Raw Data (Ca^{2+}-signals) are generally presented as photons/s or photons/s/pixel, within the desired drawn ROI. The amplitude (within the drawn ROI) is also quantified, as well as the duration of the response. Finally, with these parameters, the total amount of emitted photons of the overall response can be calculated (for a detailed example: see (39)).

4. Typical Results

4.1. Pharmacologically Induced Response in the Mushroom-Bodies

As a first step to demonstrate that the new bioluminescent GFP-aequorin (GA) Ca^{2+}-sensor was functional in the *Drosophila* brain, GA was targeted to the Mushroom-Bodies (MBs), a particular structure in the brain, intensively investigated for its role in olfactory learning and memory (61–63). As described above, using the snorkeling method, the fly is inserted and fixed, by the neck, in a blue tip. The drug is applied either manually directly within the chamber, or through the perfusion system, while a peristaltic pump allows to remove the surplus ringer. The main advantage of this approach is that the neuronal activity can be recorded continuously, without any undesirable disturbance, over long-term period (hours). This approach has allowed us to reveal that nicotine (an agonist of the ionotropic acetylcholine receptor) induced as expected, a primary response in the calyx (dendrites and the cell-bodies), but unexpectedly, a delayed secondary response exclusively in the lobes (axon/projections) (Fig. 3 in (38)). To quantify the response, Region of Interests (ROI) are drawn, and the emitted light (number of photons) is quantified, allowing us to compare several flies and/or in different conditions, e.g., different genotypes (mutants), or pharmacologically (drug) treated-flies. To visualize the Ca^{2+}-response, the accumulation time can be adjusted at our convenience (e.g., 2 s, 10 s or even 60 s), allowing to observe the response and its characteristics. Another main advantage provided by this approach is that it allows visualizing simultaneously the overall structure of the MBs, both the complex "calyx-cell-bodies," and the axonal projections (both vertical and medial lobes). Therefore, both induced Ca^{2+}-activity as well as propagation of activity in the axonal projections can be visualized and quantified simultaneously.

4.2. Neuronal Ca^{2+}-Activity Induced by Natural Stimuli (Odor-Induced Activity in ORNs)

As a second step, to validate the bioluminescent GFP-aequorin probe, and especially to demonstrate that this sensor is sensitive enough to detect neuronal Ca^{2+}-activity, experiments have also been performed using natural stimuli, like odors. Using the P(GAL4) Or83b driver to specifically target the P(UAS-GFP-aequorin) in the olfactory receptors neurons (ORNs), Ca^{2+}-responses have been recorded following successive odor applications. More precisely, progeny of flies containing both the P(GAL4)Or83b driver and the P(UAS-GFP-aequorin) transgene (Or83b,GA/Canton-S) in transheterozygotes, were used for all imaging experiments (OR83b driver targets about 80% of the ORNs (11)). In general, a single copy of the P(GAL4) driving a single copy of the UAS-GFP-aequorin is sufficient to record the odor-induced signal in the axon terminals of the ORNs (39, 40) (Note 7). Figure 3 shows a typical response of

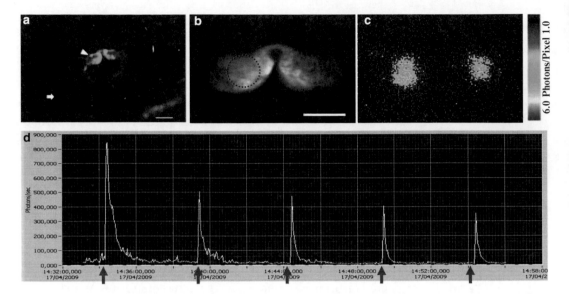

Fig. 3. A representative odor-induced Ca²⁺-responses in axon terminals of ORNs. (a) Or83b-GAL4 driver targeting the GFP-aequorin in the majority of the olfactory receptors neurons (ORNs). Combined dim-light and fluorescence images showing the ORNs in the antennae (*arrowhead*) and their synaptic terminals (*arrow*) in the antennal lobes (Leica MZ FLIII binocular, scale bar = 100 μm). (b) Fluorescence image of the antennal lobes taken at the beginning of the experiment, which will be used further as reference image. The *red-dashed circle* represents the ROI (region of interest) from which the light emission is quantified (scale bar = 50 μm). (c) A bioluminescence image (accumulation time: 10 s) of the first odor application. (d) A representative bioluminescent Ca²⁺-activity profile evoked by the application of 5 s of an odor (spearmint) (*red arrows*), five successive times, at 5 min-intervals. We remark that the Ca²⁺-response decreases (adapts) to repeated odor stimulation.

the application of 5 s of spearmint, repeated five successive times at an interval of 5 min. We can easily note that the successive responses decrease, indicating a rapid adaptation process (40). To precisely evaluate the Ca²⁺-response, three different parameters have been quantified: the amplitude (maximum height of the peak), the duration, and the total number of emitted photons (39). Moreover, especially on the first odor application, we can note that the odor-induced response presents a small shoulder, indicating that the overall Ca²⁺-response might result from the summation of successive different components with different kinetics. In other terms, it might be the superposition of at least two components. Then, by a deconvolution analysis (see (39) for the detailed deconvolution method), the second component can be revealed. This is a typical example that the absolute quantification of intracellular Ca²⁺ provided by the GA sensor is crucial to reveal such a component. In a step further, using the different genetic tools provided by *Drosophila*, an interferential RNA (RNAi) directed against a given specific gene can be expressed within the same cells than the GFP-aequorin, allowing to record its effect (knock-down of the gene) within the same neurons than the one we record the Ca²⁺-activity. In that way, targeting the inositol-1,4,5-trisphosphate receptor (InsP₃R) (P(UAS-InsP₃R-RNAi) and/or the ryanodine receptor (RyR) (P(UAS-RyR-RNAi))) within the ORNs has allowed us to demonstrate that this second component of the odor response was based on mobilization of the intracellular Ca^{2+}-stores (in the endoplasmic reticulum) (39, 40).

4.3. Deep Structures and Long-Term Recording: CC, MBs, Whole Brain Activity

As aforementioned, conversely to the different fluorescent approaches, GFP-aequorin bioluminescent approach does not require light excitation, consequently, it is not compounded by autofluorescence, phototoxicity, and photobleaching. Therefore, it allows one to record either structures that are located deep in the brain, as the central complex (CC), as well as recording over long-term periods (as long as from several hours and even few days can be performed) (Note 5). It also permits us to record the overall activity of the brain (considered as an ensemble). Different experiments have already been conducted, and few results already published (see Fig. 3 in (37)), while, for some others, preliminary data have already been obtained, validating the feasibility of these experiments.

4.3.1. Recording Deep Structures (Ellipsoïd-Body of the Central Complex)

In *Drosophila*, to date, structures that are located deep in the brain, like the CC, have not yet been accessible for the different standard physiological approaches (either electrophysiology or fluorescent Ca^{2+}-imaging), particularly because, for this last, of the undesirable autofluorescence effect. Therefore, there are very little data concerning the physiology of these deep neurons, and consequently, limiting our understanding of their neurocomputations. The CC is mainly known for its role in the control of locomotor activity (64–67). Using a P(GAL4) line (C232) (66, 68) to target the GA to the ring-neurons, a subpopulation of the ellipsoid-body neurons, Ca^{2+}-signals have been detected, following a general depolarization (KCl, 70 mM) (Fig. 2 in (38)), confirming its accessibility with the bioluminescent GFP-aequorin probe. Even, spontaneous activity (spontaneous refers to not voluntarily induced by the experimenter), occurring exclusively in the ellipsoid-body (the neuropile region) and not in the cell bodies, has been detected (unpublished results). Although we do not know yet what the fly was doing at this precise moment, it nevertheless validates that Ca^{2+}-activity can be recorded, exclusively from the synaptic connections. These preliminary results open the door to record and analyze the role of the different neurons and/or substructures of the CC in relation to the locomotor activity of the flies, particularly in conjunction to simultaneously recording the locomotor activity of the fly using the optically recorded locomotor ball (described above).

4.3.2. Long-Term Recording (e.g., Overnight in Mushroom-Bodies)

Since it is not toxic, another important advantage provided by the GFP-aequorin is that it permits recording over long-term period, up to a few days. Then, it allows recording either the induced activity or the so-called "spontaneous" activity (since we can record, in continuous, over long-term period). In such a way, overnight recording has been performed on the Mushroom-Bodies. However, to achieve such long recordings, crucial particular measures have

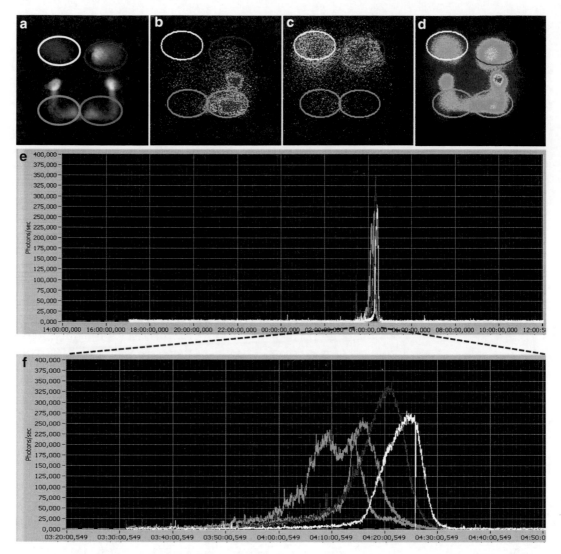

Fig. 4. Long-term recording (overnight) of spontaneous Ca²⁺-activity in the Mushroom-Bodies. (**a**) Fluorescent image taken before the experiment, and used as reference (Fly: enhancer-trap OK107-GAL4 expressing the GFP-aequorin within the Mushroom-Bodies). The four ellipsoid-circles represent the four ROI used to quantify the number of emitted photons. (**b–d**) Successive bioluminescent images (accumulation time: 30 s) showing the huge peak of spontaneous Ca²⁺-activity, mainly in the right (medial and vertical) lobes (**b**), in the calyx/cell-bodies (**c**) and in overall MBs (**d**). (**e**) Time course of the whole overnight recording, starting at 17:00 and ending at 12:00 (noon) the day after. We can see that the burst of Ca²⁺-activity occurs between 04:00 and 04:30. (**f**) Magnification of 2 h (between 03:00 and 05:00) showing the precise activity in each ROI (color-coded). We remark that activity starts in the blue ROI, which corresponds to the "right" vertical and medial lobes. Then, it progresses retrogradally, and on the route, crosses the midline (*green ROI*), while at the same time propagates to the right cell-bodies (*red-ROI*), and finally reaches the left cell-bodies (*white ROI*). See also the movie on our Web site (http://www.inaf.cnrs-gif.fr/ned/equipe10/intranet/, login: intranet-jrmartin, password = jrmartin-referee1) (covering between 03:00 and 05:00) to visualize the propagation of the neuronal Ca²⁺-activity.

to be taken (Note 5). For example, using the line OK107 to target the GFP-aequorin in the Mushroom-Bodies, the spontaneous activity occurring in the MBs has been recorded overnight (Fig. 4). While punctual activity occurs (and can be recorded) in different parts of the MBs, more interestingly, and unexpectedly, a single huge peak of Ca^{2+}-activity occurs roughly in the middle of the night (in some flies, two peaks have been observed). Looking more precisely at different successive time windows (Fig. 4b–d), we can show that this activity starts in the tip of the vertical lobes, progresses retrogradually to terminate in the cell-bodies, while on the route, crosses contralaterally, to invade the contralateral MBs. This intriguing spontaneous Ca^{2+}-activity has been revealed because this approach allows recording, in continuous, without any constraints, the spontaneous activity. This result raises several and yet unexpected questions: what is the meaning of this spontaneous activity? Since the MBs are well-known structures crucial for learning and memory (61–63) and more recently in sleep (69) one could hypothesize that this activity is involved in memory consolidation and/or in sleep. Further studies are now possible to investigate such exciting questions.

4.3.3. Recording Overall Activity of the Brain (Pan-neuronal Expression)

Again, since this approach does not require light excitation, and therefore we do not need to precisely excite, with a given wavelength light, a specific group or given neurons (like it is compulsory with the fluorescent probes), recording of the overall brain activity, considered as a whole, is possible. Using a pan-neuronal driver, like elav-Gal4 to target the GA in all neurons of the brain, overall activity (either induced or spontaneous) activity can be recorded, and so, over long-term period (for an example, see Fig. 3 in (37)). For the first time, this opens the door to trace and perform anatomical-functional mapping of the entire brain. In other words, it allows building a functional Atlas of the brain (see perspective below).

5. Conclusion/ Perspectives

5.1. Some Biological Perspectives Opened by the Bioluminescence Approach

Some new biological mechanisms have already been revealed by the bioluminescence approach, as for instance the nicotinic induced delayed secondary response in Mushroom-Bodies (38) as well as the description of an adaptation process occurring in the olfactory receptor neurons (39, 40). Other new and still unexpected biological phenomena have been observed, particularly because of the characteristics of the GFP-aequorin probe. Notably, this new approach allows continuous recording over a long time (several hours) of specific neurons in the brain, or the general activity of the brain. Indeed, targeting only few neurons, like the circadian rhythms neurons, using the pdf-gal4 driver (70), allows us to record the time when those neurons are active during the day. Preliminary (unpublished) results obtained in our lab (in collaboration with F. Rouyer's lab, CNRS, Gif-sur-Yvette, France) attest of the feasibility of those experiments. A second intriguing and exciting observation concerns the spontaneous activity in the MBs when recorded over a long period, as overnight. As cited above, we can observe a single huge burst of spontaneous Ca^{2+}-activity in the MBs in the middle of the night. Several exciting questions arise from this new observation. What is the origin of this activity? Is it cells autonomous or is it due to neuronal activity in other parts of the brain. Moreover, what is the biological meaning of this activity? All of those questions are now assessable, and are in progress in our lab.

Another exciting perspective opened by this new approach is the possibility, for the first time, to produce an anatomical-functional map of the general and overall activity of the brain, for a short or long time period. Indeed, using a pan-neural driver, like P(elav-GAL4), or an ubiquitous driver like P(daughterless-GAL4) to express the GFP-aequorin, will allow us to record as an ensemble, the general activity of the brain (either evoked or spontaneous activity), simultaneously in all "active" neurons. Then this could lead to production of a functional map (or in other term, a functional ATLAS) of the activity of the brain (similar to and in complement to the static (morphological) ATLASes of the brain made by anatomists). Those maps would be a prerequisite to further investigate other conditions, such as the difference between males and females (sexual dimorphism) or the age of the fly (*Drosophila* is a good model of aging and longevity) or in any pharmacological context. For example, how is the general brain activity modified by the application of a drug, e.g., narcoleptic, anxiolytic, psychostimulant? Additionally, *Drosophila* models have recently been developed to study the neurophysiological mechanisms of the addiction to certain drugs (ethanol, cocaine, nicotine: for review, see (71, 72)). Thus, how is the overall general brain activity of a "drunk" (ethanol intoxicated) fly?

Preliminary data show strong promise for the feasibility of this long-term recording (37). In a similar manner, using repo-GAL4 to express GFP-aequorin in the glial cells (or any other specific glial cells drivers) will also allow to record Ca^{2+}-activity in those cells. Also here, preliminary data demonstrate that glial cell Ca^{2+}-activity could be recorded over a long time period (overnight). Another possibility could be to perform those maps on brain affected by a given neuropathology. Indeed, different *Drosophila* models exist for several neurodegenerative human diseases as Parkinson's, Alzheimer's, and Huntington (for reviews, see (73, 74)). Our current knowledge about the modifications of the general brain activity in relation with those pathologies is still limited. Similarly, a map of functional activity on brain that harbors a mutation (or two or more) in a specific and well characterized gene (s) will also be feasible. For those purposes, *Drosophila* is an ideal organism, since several hundred mutants are already well molecularly and cellularly characterized, as for example the gene affecting the cAMP signaling pathway (*dunce, rutabaga, amnesiac*) (62) or *caki* mutant flies (*caki* encodes for a MAGUK (membrane-associated guanylate kinase) protein) (75). Nevertheless, for the moment, since we can record solely from above, the map can be traced only in 2D (3D with time). However, recording simultaneously from two different angles (ideally from 90°) would allow a 3D reconstruction of the image (4D with time) and therefore would allow us to produce a complete 3D Atlas of the brain (4D including time).

5.2. Some Perspectives in Technological Improvements

5.2.1. Development of Bioluminescent Probes with Different Wavelength Emission

The emission wavelength is a characteristic of each natural bioluminescent probe. Indeed, several other bioluminescent molecules have been identified, as for example obelin (76) and can potentially be used to perform in vivo imaging. Moreover, like the GFP-aequorin, they are likely usable, as a starting point, to be genetically engineered, with the goal to modify and improve them. In such a way, two variants of the GFP-aequorin, the YFP-aequorin (Venus) and the RFP-aequorin have been developed (35). The light-emission of Venus is shifted in the yellow, while, in RFP-aequorin, based on the RFP of the red coral, the light emission is shifted in the red, and even partly, in the infrared. The main goal of developing new sensor-probes, and particularly for the red emitting bioluminescent molecules is that red or infrared light crosses the soft tissues more efficiently, with less attenuation and scattering than the green or yellow (visible) light (77). This could yield to generate a Ca²⁺-sensor that could potentially monitor the neuronal Ca²⁺-activity without the need to open the brain capsule (for invertebrates). This would allow detecting the bioluminescence emitted light, in the intact animal, even potentially in the free moving animal. Another advantage of disposing of an alternative sensor emitting in a different wavelength, would permit to use it simultaneously in the same fly, with the GFP-aequorin, allowing monitoring simultaneously two different cell populations. Indeed, with the genetic tools of *Drosophila*, using either a second binary system, or a specific DNA promoter region, we could express in the same fly, different probes, as for example, the RFP-aequorin (red) in the glial cells and the GFP-aequorin (green) in the neuronal cells. This would permit us to monitor functionally and concomitantly the activity of those two cell populations and study their interaction. Alternatively, red-emitting (RFP) and green-emitting (GFP-aequorin) could also be used to simultaneously record Ca²⁺-signals from two different organelles in the same cell, as for instance, in the nucleus or the mitochondria versus the cytoplasm. Such experiments have already been done in cell cultures (78). However, to perform such bi-recording, we will need to improve and implement the recording setup with a second camera, to record the red (and infrared) light, simultaneously with the green one. This will be feasible using a splitting dichroic mirror, in which the emitted light will be separated into two beams.

Improvement of the bioluminescence detecting setup are also currently under development. Particularly, up to now, the optical system used (the objectives) have been designed to the needs and constraints of the fluorescence. Notably the objectives are composed of several corrective lenses to correct the image. Consequently, the brightness of the objective is reduced (which is generally not, in this range of light, a limiting factor for the fluorescence). However, since in bioluminescence the main limiting factor is the amount of light, the brightness of the objective could become a crucial and limiting parameter. To date, by default, we currently use the one dedicated to the fluorescence. Undoubtedly, the development of more appropriate objectives with a higher brightness is desirable. Another crucial component of the setup is the photon detector and CCD camera, and particularly its sensitivity. Other photon detectors and CCD cameras are currently under development, which might allow a more sensitive detection, and consequently would also permit a better temporal resolution. For example, pilot experiments have shown that using an ICC camera (Stanford Photonics, USA), we could obtain a better sensitivity, and consequently the temporal resolution could be increased (up to 10 ms). Even a temporal resolution down to 1 ms (1,000 Hz) seems to be feasible. With this improved sensitivity, preliminary data performed on *Drosophila* have shown that Ca^{2+}-signals can be detected in the antennae through the cuticle, without opening.

In conclusion, several avenues are currently under development to improve the bioluminescence approach, both at the molecular level, to improve the light emitted by the probe, or at the technological level, by improving the setup used to detect the light. Several preliminary experiments have shown that numerous parameters can still be improved, which will allow important improvement in the general approach. Undoubtedly, we reckon that this new and still young brain imaging technique is extremely promising and should surely bring important breakthroughs and advances in our understanding of the functioning of the brain.

6. Notes (Troubleshooting)

1. Different types of coelenterazine. Several types (up to 37) of chemically modified analogues of coelenterazine have been synthesized (e.g., cp, e, f, fcp, h, hcp, etc.) and are commercially available (e.g., NanoLight, Uptima, Molecular Probes, etc.), with a large variability in the price (cost). Table 1 (23) summarizes the different coelenterazine with their different maximum emission wavelength, relative luminescence capacity, relative intensity, and half-rise time. To date, for the functional in vivo brain imaging, we have used the native (*n*-coelenterazine)

or the benzyl (*h*-coelenterazine). Moreover, it should also be noticed that for the recording of a deeper tissue, it is preferable to incubate longer (≥ 2 h).

2. Comparison (advantages/disadvantages) between the Photon Detector and the EMCCD camera. Though the precise comparison between the two devices (Photon Detector and EMCCD camera) is not always straightforward, some comparisons can still be made. First, the Photon Detector has the tendency to be more sensitive, and can have, in principle, a better temporal resolution (in the range of ms and even µs). This is particularly due because the EMCCD camera has a reading noise (white noise induced each time that the camera is read and resettled). However, since the temporal resolution is limited by some software/hardware constraints, up to now, the real and used temporal resolution was almost equivalent between the two systems. Second, the Photon Detector has a relatively narrower band of sensitivity to the light (e.g., in the blue and green), while the EMCCD camera has a very large spectral sensitivity (from infrared up to ultraviolet). Third, the spatial resolution (the size of the pixel) is generally better with the EMCCD camera. Fourth, concerning the cost of the devices, the Photon Detector is generally much more expensive than the EMCDD camera (about the double of price), while the Photon Detector is very fragile (the light detector can be easily burned (damage) while the EMCCD is robust and can support unexpected high amount of light). In brief, the EMCCD camera seems to be more flexible in terms of evolution of the needs, and it is more accessible (cost). Nevertheless, in certain conditions, if a high sensitivity is required, the Photon Detector will be most appropriate.

3. Numerical aperture of the objective. The relatively low level of light emitted by the bioluminescent probe GFP-aequorin remains an important limiting factor (constraint) in the light detection. Consequently, it is highly recommended to use an objective with the highest (maximum) numerical aperture (N/A) available by the company/trade of the used microscope. The light transmitted by the objective is given by the formula: $N/A \times$ Magnification. For example, an objective of 20× with a N/A of 0.75 will transmit fourfold more light than a 20× with a 0.5 N/A. Then, ideally, a objective with a N/A of 0.75 will be used, and when possible, a dry objective, because generally, the water immersion objectives have lower N/A than dry objective. The production of objectives designed and dedicated specifically to the bioluminescence is also desirable, and hopefully will be built by companies in a near future.

4. Fly preparation. It is highly recommended to not anesthetize the fly with CO_2, because it seems to cause long-term (irreversible)

damage (79). Indeed, among different measured neurophysiological/neuroethological parameters, the general locomotor activity is importantly decreased for few hours after CO_2 anesthesia (80), suggesting that CO_2 causes strong detrimental effect on the general neuronal activity. Though chilling also yields some detrimental defects, ice anesthesia seems to be less severe than CO_2. Consequently, it is highly recommended to minimize as maximum the time of anesthesia during the fly preparation. The choice of the glue to fix the fly is also very important. Non toxic glue should be used. Avoid any glue containing cyanoacrylate (like Super Glue or Crazy glue™). Silicone needs also to be carefully selected. For example, two components silicone "biologically compatible" (e.g., Kwik-Sil™, World Precision Instruments, USA) is recommended.

5. Long-term imaging period. Because the GFP-aequorin does not require light excitation and is not toxic, long-term imaging period, as long as overnight and even longer, can be performed. However, for such long recording, the limiting factor is essentially to keep the fly alive, and avoid its dehydration. To circumvent that, the fly has to be fed during the recording, by allowing it to drink sugar-water. A small Eppendorf filled with 5% glucose in the water is attached on the holder of the fly, and a tiny capillary paper brings the sugar-water to the proboscis of the fly. Second, to avoid that the fly dehydrates (since ventilation is continuously switched-on in the black dark-box), a humid chamber surrounding the fly preparation and the objective is made. This can be hand-made by using a water-imbedded sponge placed under the holder of the fly, while the overall (the sponge, the fly and the objective) are surrounded by a plastic cone and sealed with parafilm™. In such a way, up to now, a fly could live easily for 24 h, and even in the best case, up to 48 h.

6. Temporal resolution. Currently, mainly due to some constraints in the software/hardware interface, the maximum data acquisition is 50 ms. However, new software version linked to some technical improvement in the photon detection, currently under development, should soon allow reaching a temporal resolution up to 1 ms. Indeed, because of its fast light emission following Ca^{2+}-binding (~5 ms) (33), the GFP-aequorin allows such fast imaging. Interestingly, at 1 ms, the bioluminescent recording will be in the same temporal range than the electrophysiological recording.

7. Number of copy of the P(GAL4) driver line and of the reporter P(UAS-GFP-aequorin). The number of copy of the GAL4 driver and of the UAS-reporter is also a crucial parameter. For example, up to now, in the olfactory receptor neurons (ORNs), the use of one copy of the GAL4 driver (OR83b-GAL4) and

one copy of the reporter (UAS-GFP-aequorin) (both in transhet-erozygous) has revealed to be sufficient to easily observe and record the neuronal activity at the axon terminals of the ORNs in the antennal lobes (39, 40). However, to record the Projection Neurons (PNs), for which we use the GH146 driver (which is a relatively weak driver), two copies of the GAL4 driver and at least two copies of the GFP-aequorin (both in homozygous) are required to detect a reliable and quantifiable signal (unpublished results). This is likely due to two main reasons: first, the GH146 driver is relatively a weak driver that targets just a limited number of neurons (about 50 neurons). Second, in the antennal lobes, the PNs are the post-synaptic neurons, and therefore the neuronal part which is recorded corresponds to the dendrites of the neurons (which are relatively smaller than the axon terminals). Finally, since the level of the bioluminescent Ca²⁺-signal is generally dependent of the amount of probe expressed in the neurons, care might be taken to compare the different flies recorded in the same genetic conditions (same copy's number of driver and/or reporter gene).

Acknowledgments

I'm indebted to P. Brûlet and his colleagues, K. Rogers, and S. Picaud, who have primarily developed the bioluminescence approach. I also thank the different past and present members of my laboratory (students and Post-Docs: E. Carbognin, M.S. Murmu, E. Real, P. Pavot, A. Khammari and L. Mellottée) who have participated in the development of the bioluminescence approach or recording neuronal activity, in *Drosophila*. I also thanks D. Nässel for the critical reading of the manuscript, as well as the French "Agence National pour la Recherche" (ANR-Neuroscience), the NERF (Neuropôle Ile-de France), and the CNRS, for their precious financial support.

References

1. Miyawaki A (2005) Innovations in the imaging of brain functions using fluorescent proteins. Neuron 48:189–199

2. Griesbeck O (2004) Fluorescent proteins as sensors for cellular functions. Curr Opin Neurobiol 14:636–641

3. Garaschuk O, Milos RI, Grienberger C, Marandi N, Adelsberger H, Konnerth A (2006) Optical monitoring of brain function in vivo: from neurons to networks. Pflugers Arch 453:385–396

4. Dombeck DA, Khabbaz AN, Collman F, Adelman TL, Tank DW (2007) Imaging large-scale neural activity with cellular resolution in awake, mobile mice. Neuron 56:43–57

5. Kerr R, Lev-Ram V, Baird G, Vincent P, Tsien RY, Schafer WR (2000) Optical imaging of calcium transients in neurons and pharyngeal muscle of *C. elegans*. Neuron 26:583–594

6. Fiala A, Spall T, Diegelmann S, Eisermann B, Sachse S, Devaud JM, Buchner E, Galizia CG (2002) Genetically expressed cameleon in

Drosophila melanogaster is used to visualize olfactory information in projection neurons. Curr Biol 12:1877–1884

7. Fiala A, Spall T (2003) *In vivo* calcium imaging of brain activity in *Drosophila* by transgenic cameleon expression. Sci STKE 174:PL6

8. Nakai J, Ohkura M, Imoto K (2001) A high signal-to-noise Ca²⁺ probe composed of a single green fluorescent protein. Nat Biotechnol 19:137–141

9. Wang Y, Wright NJ, Guo H, Xie Z, Svoboda K, Malinow R, Smith DP, Zhong Y (2001) Genetic manipulation of the odor-evoked distributed neural activity in the *Drosophila* mushroom body. Neuron 29:267–276

10. Wang Y, Guo HF, Pologruto TA, Hannan F, Hakker I, Svoboda K, Zhong Y (2004) Stereotyped odor-evoked activity in the mushroom body of *Drosophila* revealed by green fluorescent protein-based Ca²⁺ imaging. J Neurosci 24:6507–6514

11. Wang JW, Wong AM, Flores J, Vosshall LB, Axel R (2003) Two-photon calcium imaging reveals an odor-evoked map of activity in the fly brain. Cell 112:271–282

12. Yu D, Baird GS, Tsien RY, Davis RL (2003) Detection of calcium transients in *Drosophila* mushroom body neurons with camgaroo reporters. J Neurosci 23:64–72

13. Yu D, Ponomarev A, Davis RL (2004) Altered representation of the spatial code for odors after olfactory classical conditioning; memory trace formation by synaptic recruitment. Neuron 42:437–449

14. Yu D, Keene AC, Srivatsan A, Waddell S, Davis RL (2005) Drosophila DPM neurons form a delayed and branch-specific memory trace after olfactory classical conditioning. Cell 123:945–957

15. Yu D, Akalal DB, Davis RL (2006) Drosophila alpha/beta mushroom body neurons form a branch-specific, long-term cellular memory trace after spaced olfactory conditioning. Neuron 52:845–855

16. Higashijima S, Masino MA, Mandel G, Fetcho JR (2003) Imaging neuronal activity during zebrafish behavior with a genetically encoded calcium indicator. J Neurophysiol 90:3986–3997

17. Hara M, Bindokas V, Lopez JP, Kaihara K, Landa LR Jr, Harbeck M, Roe MW (2004) Imaging endoplasmic reticulum calcium with a fluorescent biosensor in transgenic mice. Am J Physiol Cell Physiol 287:C932–C938

18. Hasan MT, Friedrich RW, Euler T, Larkum ME, Giese G, Both M, Duebel J, Waters J, Bujard H, Griesbeck O, Tsien RY, Nagai T, Miyawaki A, Denk W (2004) Functional fluorescent Ca²⁺ indicator proteins in transgenic mice under TET control. PLoS Biol 2:e163

19. Nagai T, Yamada S, Tominaga T, Ichikawa M, Miyawaki A (2004) Expanded dynamic range of fluorescent indicators for Ca²⁺ by circularly permuted yellow fluorescent proteins. Proc Natl Acad Sci USA 101:10554–10559

20. Vincent P, Maskos U, Charvet I, Bourgeais L, Stoppini L, Leresche N, Changeux JP, Lambert R, Meda P, Paupardin-Tritsch D (2006) Live imaging of neural structure and function by fibred fluorescence microscopy. EMBO Rep 11:1154–1161

21. Reiff DF, Ihring A, Guerrero G, Isacoff EY, Joesch M, Nakai J, Borst A (2005) *In vivo* performance of genetically encoded indicators of neural activity in flies. J Neurosci 25:4766–4778

22. Shimomura O, Johnson FH (1978) Peroxidized coelenterazine, the active group in the photoprotein aequorin. Proc Natl Acad Sci USA 75:2611–2615

23. Shimomura O, Musicki B, Kishi Y (1989) Semi-synthetic aequorins with improved sensitivity to Ca²⁺ ions. Biochem J 261:913–920

24. Shimomura O, Musicki B, Kishi Y, Inouye S (1993) Light-emitting properties of recombinant semi-synthetic aequorins and recombinant fluorescein-conjugated aequorin for measuring cellular calcium. Cell Calcium 14:373–378

25. Leclerc C, Webb SE, Daguzan C, Moreau M, Miller AL (2000) Imaging patterns of calcium transients during neural induction in Xenopus laevis embryos. J Cell Sci 113:3519–3529

26. Gilland E, Miller AL, Karplus E, Baker R, Webb SE (1999) Imaging of multicellular large-scale rhythmic calcium waves during zebrafish gastrulation. Proc Natl Acad Sci USA 96:157–161

27. Rosay P, Davies SA, Yu Y, Sözen A, Kaiser K, Dow JA (1997) Cell-type specific calcium signalling in a *Drosophila* epithelium. J Cell Sci 110:1683–1692

28. Rosay P, Armstrong JD, Wang Z, Kaiser K (2001) Synchronized neural activity in the *Drosophila* memory centers and its modulation by amnesiac. Neuron 30:759–770

29. Torfs H, Poels J, Detheux M, Dupriez V, Van Loy T, Vercammen L, Vassart G, Parmentier M, Vanden Broeck J (2002) Recombinant aequorin as a reporter for receptor-mediated changes of intracellular Ca²⁺-levels in *Drosophila* S2 cells. Invert Neurosci 4:119–124

30. Kerr M, Davies SA, Dow JA (2004) Cell-specific manipulation of second messengers; a toolbox for integrative physiology in *Drosophila*. Curr Biol 14:1468–1474

31. MacPherson MR, Pollock VP, Kean L, Southall TD, Giannakou ME, Broderick KE, Dow JA, Hardie RC, Davies SA (2005) Transient receptor potential-like channels are essential for calcium signaling and fluid transport in a *Drosophila* epithelium. Genetics 169:541–1552

32. Baubet V, Le Mouellic H, Campbell AK, Lucas-Meunier E, Fossier P, Brulet P (2000) Chimeric green fluorescent protein-aequorin as bioluminescent Ca²⁺ reporters at the single-cell level. Proc Natl Acad Sci USA 97:7260–7265

33. Gorokhovatsky AY, Marchenkov VV, Rudenko NV, Ivashina TV, Ksenzenko VN, Burkhardt N, Semisotnov GV, Vinokurov LM, Alakhov YB (2004) Fusion of *Aequorea victoria* GFP and aequorin provides their Ca²⁺-induced interaction that results in red shift of GFP absorption and efficient bioluminescence energy transfer. Biochem Biophys Res Commun 320:703–711

34. Rogers KL, Stinnakre J, Agulhon C, Jublot D, Shorte SL, Kremer EJ, Brulet P (2005) Visualization of local Ca²⁺ dynamics with genetically encoded bioluminescent reporters. Eur J Neurosci 3:597–610

35. Curie T, Rogers KL, Colasante C, Brûlet P (2007) Red-shifted aequorin-based bioluminescent reporters for *in vivo* imaging of Ca²⁺ signaling. Mol Imaging 6:30–42

36. Rogers KL, Picaud S, Roncali E, Boisgard R, Colasante C, Stinnakre J, Tavitian B, Brûlet P (2007) Non-invasive *in vivo* imaging of calcium signaling in mice. PLoS One 2:e974

37. Rogers KL, Martin JR, Renaud O, Karplus E, Nicola MA, Nguyen M, Picaud S, Shorte SL, Brûlet P (2008) EMCCD based bioluminescence recording of single-cell Ca²⁺. J Biomed Opt 13:1–10

38. Martin JR, Rogers KL, Chagneau C, Brûlet P (2007) *In vivo* bioluminescence imaging of Ca²⁺ signalling in the brain of *Drosophila*. PLoS One 2:e275

39. Murmu MS, Stinnakre J, Martin JR (2010) Presynaptic Ca²⁺-stores contribute to odor-induced response in Drosophila olfactory receptor neurons. J Exp Biol 213:4163–4173

40. Murmu MS, Stinnakre J, Réal E, Martin JR (2011) Calcium-stores mediate adaptation in axon terminals of Olfactory Receptor Neurons in Drosophila. BMC Neurosci 12:105

41. Adams MD et al (2000) The genome sequence of Drosophila melanogaster. Science 287:2185–2195

42. Brand AH, Perrimon N (1993) Targeted gene expression as a means of altering cell fates and generating dominant phenotypes. Development 118:401–415

43. Elliott DA, Brand AH (2008) The GAL4 system: a versatile system for the expression of genes. Methods Mol Biol 420:79–95

44. Lai SL, Lee T (2006) Genetic mosaic with dual binary transcriptional systems in Drosophila. Nat Neurosci 9:703–709

45. Potter CJ, Tasic B, Russler EV, Liang L, Luo L (2010) The Q system: a repressible binary system for transgene expression, lineage tracing, and mosaic analysis. Cell 141:536–548

46. Sweeney ST, Broadie K, Keane J, Niemann H, O'Kane CJ (1995) Targeted expression of tetanus toxin light chain in *Drosophila* specifically eliminates synaptic transmission and causes behavioral defects. Neuron 14:341–351

47. Martin JR, Keller A, Sweeney ST (2002) Targeted expression of tetanus toxin: a new tool to study the neurobiology of behavior. Adv Genet 47:1–47

48. Kitamoto T (2001) Conditional modification of behavior in *Drosophila* by targeted expression of a temperature-sensitive *shibire* allele in defined neurons. J Neurobiol 47:81–92

49. Pulver SR, Pashkovski SL, Hornstein NJ, Garrity PA, Griffith LC (2009) Temporal dynamics of neuronal activation by Channelrhodopsin-2 and TRPA1 determine behavioral output in Drosophila larvae. J Neurophysiol 101:3075–3088

50. Peabody NC, Pohl JB, Diao F, Vreede AP, Sandstrom DJ, Wang H, Zelensky PK, White BH (2009) Characterization of the decision network for wing expansion in *Drosophila* using targeted expression of the TRPM8 channel. J Neurosci 29:3343–3353

51. White B, Osterwalder T, Keshishian H (2001) Molecular genetic approaches to the targeted suppression of neuronal activity. Curr Biol 11:R1041–R1053

52. Hodge JJ (2009) Ion channels to inactivate neurons in *Drosophila*. Front Mol Neurosci 2:13

53. Aso Y, Grubel K, Busch S, Friedrich AB, Siwanowicz I, Tanimoto H (2009) The mushroom body of adult Drosophila characterized by GAL4 drivers. J Neurogenet 23:156–172

54. Ashburner M (1989) *Drosophila*, A laboratory manual. Cold Spring Harbor Laboratory Press, Cold Spring Harbor, NY

55. Roberts DB (1998) *Drosophila*, a practical approach. Oxford University Press, Oxford

56. Gu H, O'Dowd DK (2006) Cholinergic synaptic transmission in adult Drosophila Kenyon cells in situ. J Neurosci 26:265–272

57. Agulhon C, Platel JC, Kolomiets B, Forster V, Picaud S, Brocard J, Faure P, Brulet P (2007) Bioluminescent imaging of Ca²⁺ activity reveals

spatiotemporal dynamics in glial networks of dark-adapted mouse retina. J Physiol 583: 945–958

58. Miller AL, Karplus E, Jaffe LF (1994) Imaging $(Ca^{2+})_i$ with aequorin using a photon imaging detector. Methods Cell Biol 40:305–338

59. Kazama H, Wilson RI (2008) Homeostatic matching and nonlinear amplification at identified central synapses. Neuron 58:401–413

60. Yaksi E, Wilson RI (2010) Electrical coupling between olfactory glomeruli. Neuron 67: 1034–1047

61. Heisenberg M (2003) Mushroom body memoir: from maps to models. Nat Rev Neurosci 4:266–275

62. Davis RL (2005) Olfactory memory formation in *Drosophila*: from molecular to systems neuroscience. Annu Rev Neurosci 28:275–302

63. Davis RL (2011) Traces of Drosophila memory. Neuron 70:8–19

64. Strauss R, Heisenberg M (1993) A higher control center of locomotor behavior in the *Drosophila* brain. J Neurosci 13:1852–1861

65. Strauss R (2002) The central complex and the genetic dissection of locomotor behaviour. Curr Opin Neurobiol 12:633–638

66. Martin JR, Raabe T, Heisenberg M (1999) Central complex substructures are required for the maintenance of locomotor activity in *Drosophila melanogaster*. J Comp Physiol A 185:277–288

67. Martin JR, Faure F, Ernst R (2002) The power law distribution for walking-time intervals correlates with the ellipsoid-body in *Drosophila*. J Neurogenet 15:1–15

68. Renn SC, Armstrong JD, Yang M, Wang Z, An X, Kaiser K, Taghert PH (1999) Genetic analysis of the Drosophila ellipsoid body neuropil: organization and development of the central complex. J Neurobiol 41:189–207

69. Joiner WJ, Crocker A, White BH, Sehgal A (2006) Sleep in Drosophila is regulated by adult mushroom bodies. Nature 441:757–760

70. Renn SC, Park JH, Rosbash M, Hall JC, Taghert PH (1999) A pdf neuropeptide gene mutation and ablation of PDF neurons each cause severe abnormalities of behavioral circadian rhythms in *Drosophila*. Cell 99:791–802

71. Bellen HJ (1998) The fruit fly: a model organism to study the genetics of alcohol abuse and addiction? Cell 93:909–912

72. Wolf FW, Heberlein U (2003) Invertebrate models of drug abuse. J Neurobiol 54: 161–178

73. Bilen J, Bonini NM (2005) *Drosophila* as a model for human neurodegenerative disease. Annu Rev Genet 39:153–171

74. Muqit MM, Feany MB (2002) Modelling neurodegenerative diseases in *Drosophila*: a fruitful approach? Nat Rev Neurosci 3:237–243

75. Martin JR, Ollo R (1996) A new Drosophila Ca^{2+}/calmodulin-dependent protein kinase (Caki) is localized in the central nervous system and implicated in walking speed. EMBO J 15:1865–1876

76. Markova SV, Vysotski ES, Blinks JR, Burakova LP, Wang BC, Lee J (2002) Obelin from the bioluminescent marine hydroid Obelia geniculata: cloning, expression, and comparison of some properties with those of other Ca^{2+}-regulated photoproteins. Biochemistry 41:2227–2236

77. Bakayan A, Vaquero CF, Picazo F, Llopis J (2011) Red fluorescent protein-aequorin fusions as improved bioluminescent Ca^{2+} reporters in single cells and mice. PLoS One 6(5):e19520

78. Manjarrés IM, Chamero P, Domingo B, Molina F, Llopis J, Alonso MT, García-Sancho J (2008) Red and green aequorins for simultaneous monitoring of Ca^{2+} signals from two different organelles. Pflugers Arch 455:961–970

79. Barron AB (2000) Anaesthetising Drosophila for behavioural studies. J Insect Physiol 2000(46):439–442

80. Martin JR (2003) Locomotor activity: a complex behavioural trait to unravel. Behav Processess 64:145–160

Chapter 2

Ca²⁺ Imaging in Brain Slices Using Bioluminescent Reporters

Ludovic Tricoire, Estelle Drobac, and Bertrand Lambolez

Abstract

Imaging of Ca^{2+} indicators is widely used to record transient intracellular Ca^{2+} increases associated with bioelectrical activity. The natural bioluminescent Ca^{2+} sensor aequorin has been historically the first Ca^{2+} indicator used to address biological questions. Aequorin is generally superseded today by fluorescent Ca^{2+} indicators for imaging applications that require high spatial and temporal resolution. Nonetheless, aequorin imaging offers several advantages over fluorescent reporters: it is virtually devoid of background signal; it does not require light excitation and interferes little with intracellular processes. In this chapter, we describe protocols allowing the expression of a GFP-aequorin fusion protein in acute brain slices and the bioluminescence recording of Ca^{2+} transients in single neurons, or multiple neurons simultaneously.

Key words: Ca^{2+} imaging, Bioluminescence, Aequorin, Photoproteins, Viral gene transfer, Neuron, Neuronal network

1. Introduction

Aequorin, isolated from jellyfish *Aequorea* species, is a bioluminescent complex that emits blue light upon Ca^{2+} binding (1). Aequorin was first used as a Ca^{2+} indicator to evidence the role of intracellular Ca^{2+} in excitation–contraction coupling (2), in neuronal signaling (3, 4) and in meiosis (5). Since then, aequorin has been extensively used as a reporter of Ca^{2+} physiology in various cell-types and subcellular compartments following intracellular injection or expression by gene transfer (reviewed in (6)).

Aequorin is a stable luciferase intermediate formed from the reaction of the protein apoaequorin (luciferase) and the prosthetic group coelenterazine (luciferin), and contains three EF-hand Ca^{2+}-binding sites (7–9). The formation of the aequorin complex is a slow process, whereas the bioluminescence reaction occurs as a rapid flash whose intensity increases with Ca^{2+} concentration, and proceeds to completion in the continuous presence of Ca^{2+} (7, 10–12).

Jean-René Martin (ed.), *Genetically Encoded Functional Indicators*, Neuromethods, vol. 72,
DOI 10.1007/978-1-62703-014-4_2, © Springer Science+Business Media, LLC 2012

The rapid kinetics of Ca^{2+} binding to and unbinding from aequorin makes it a suitable indicator of rapid Ca^{2+} transients (10). Aequorin belongs to a family of Ca^{2+} sensors named photoproteins, whose members share high structural and functional homology (13). The photoprotein obelin is well characterized and also suitable for bioluminescence imaging of intracellular Ca^{2+} transients (14–18). Aequorin and Obelin exhibit almost identical Ca^{2+} sensitivity, which span Ca^{2+} concentrations from 0.1 µM to 1 mM, but differ in their kinetic properties and their sensitivity to Mg^{2+} (12, 15–17, 19). In addition, aequorin mutants and semi-synthetic aequorins (i.e., incorporating synthetic coelenterazine analog) endowed with different kinetics or Ca^{2+} sensitivity (12, 20–22) expand the range of bioluminescent Ca^{2+} sensors available to diverse applications (23–25).

A GFP-aequorin (GA) fusion protein has been described that allows fluorescence labeling of expressing cells and bioluminescence Ca^{2+} imaging of single cultured cells, of tissue slices and in whole vertebrate and invertebrate animals (17, 26–30). The fact that no illumination is required has allowed the use of GA to report neuronal activity in freely behaving zebrafishes (30). No conspicuous morphological or functional alteration has been observed upon acute high-level expression of GA or in transgenic animals stably expressing GA (27, 28, 30). Its bioluminescence in cellulo is superior to that of aequorin alone, presumably because the latter is rapidly degraded in the cytoplasm (31). The equivalent GFP-Obelin fusion protein (GO) exhibits a bioluminescence intensity similar to that of GA and the Ca^{2+} sensing properties of both fusion proteins are very close to those of the photoproteins alone (17, 26). Bioluminescence resonance energy transfer between the photoprotein and GFP moieties of both fusion proteins shifts light emission to the green (26, 32) as observed between native fluorescent proteins and photoproteins in light emitting cells of *Aequorea* and *Obelia* species (33).

Bioluminescence imaging with photoproteins is endowed with a high signal-to-noise ratio. Indeed, photoproteins exhibit negligible Ca^{2+}-independent luminescence and their response intensity vary by several orders of magnitude depending on Ca^{2+} concentration (15, 16, 19). GFP-photoproteins behave as low affinity, supralinear indicators of Ca^{2+} transients associated with action potentials in mammalian neurons (17). A detection threshold of five action potentials has been reported in cortical neurons upon bioluminescence imaging in brain slices, corresponding to Ca^{2+} concentration transients locally reaching the micromolar range (17). This detection threshold is comparable to that reported for the GCaMP genetically encoded fluorescent Ca^{2+} sensor (34). The substrate coelenterazine is generally loaded once for reconstitution of active photoprotein prior to imaging experiments. The absence of coelenterazine during the course of the experiment eventually leads to exhaustion of the active photoprotein. However, due to the relatively small percentage of aequorin consumed, long duration continuous recordings of several hours have been reported (30).

In the next sections, we describe protocols routinely used in our laboratory to perform bioluminescence imaging of Ca^{2+} transients in single neurons or from a large number of neurons simultaneously in acute rodent brain slices using GA or GO. Although other viral vectors such as adenovirus can be used to express these reporters (29), we use sindbis vector to transduce acute cortical slices because it allows fast expression (with 6–8 h) of the transgene at a high level selectively into neurons (35–37).

2. Materials

2.1. Recombinant Sindbis Vector

1. Plasmids: pSinRep5 and the helper pDH(26S) (38)
2. In vitro SP6 transcription kit (e.g., SP6 Megascript kit from Ambion)
3. Cap Analog (m7G(5′)ppp(5′)G, e.g., from Ambion)
4. Proteinase K (stock solution at 20 mg/ml)
5. Sodium dodecylsulfate (SDS, stock solution at 10%)
6. Phenol–chloroform–isoamyl alcohol solution
7. RNAse inhibitor (e.g., recombinant RNAsin from Promega)
8. Baby hamster kidney-21 cells (BHK, available from ATCC, No CCL-10)
9. Dulbecco modified Earle medium (DMEM)
10. Fetal calf serum (FCS)
11. Electroporator with cuvette 0.4 cm (e.g., Gene Pulser II with capacitance extender from Biorad)

Caution: The low level of pathogenicity of Sindbis virus in humans has allowed it to be classified as a Biosafety Level-2 (BL-2) agent by the NIH Recombinant DNA Advisory Committee.

2.2. Preparation of Neocortical Slices

1. Acute neocortical slices.
2. Minimum essential medium (MEM, e.g., from Invitrogen).
3. Hank's balanced salt solution (HBSS, e.g., from Invitrogen).
4. Glucose.
5. Penicillin and streptomycin.
6. Organotypic insert (e.g., Millicell-CM from Millipore).
7. Native coelenterazine free base (Prolume, Cat# 303 NF-CTZ-FB), stock solution at 1.25 mM in ethanol. Stored at −80°C. Caution: manipulate coelenterazine in the dark.

2.3. Imaging

Wide-field bioluminescence imaging is performed using an intensified CCD video camera (ICCD225; 768 × 576 pixels; Photek, St Leonards on Sea, UK) mounted on the C-mount port of an upright BX51WI microscope (Olympus) and controlled by the data acquisition software IFS32 (Photek). Light acquisition is performed through water immersion objectives 10× (N.A. = 0.3) and 60× (N.A. = 0.9). The imaging setup (see Fig. 1) is housed in a dark box to avoid light noise. Slices are set in a recording chamber that is continuously perfused with standard artificial cerebrosplinal fluid (ACSF). GFP fluorescence is visualized using a mercury lamp with a standard GFP filter set (e.g., from Chroma or Semrock). Filters are removed from light path during bioluminescence recordings.

3. Methods

3.1. Production of Sindbis Virus Expressing GFP-Aequorin

GFP-aequorin (GA; Genbank accession number: EF212028) and GFP-obelin (GO) consist of a GFP-coding sequence followed by the sequence coding for apoaequorin and apoobelin respectively. The two coding sequences are separated by a five-repeat linker sequence $(SGGSGSGGQ)_5$. The pattern of our fusion constructs is similar to that of the G5A construct described previously by Baubet et al. (26). The GFP coding sequence (kind gift of Jonathan Gilthorpe) is derived from EGFP (Clontech, Saint-Germain-en-Laye, France) and contained mutations F64L, S65T, N144D, V163A, I167T, S175G, and H231L. The apoaequorin gene is fully codon-optimized for better expression in mammalian cells. The pSinRep5 plasmids containing GA, as well as the helper plasmid pDH26S (Invitrogen), are linearized using PacI and XhoI respectively. After proteinase K/SDS treatment, phenol/chloroform extraction and ethanol precipitation, plasmids are then transcribed in vitro into capped RNA using the Megascript SP6 kit. BHK-21 cells are electroporated with GA and the helper RNAs, $(20 \times 10^6$ cells/ml, 950 µF, 230 V) and grown 24 h at 37°C, 5% CO_2 in DMEM containing 5% FCS before collecting the cell supernatant 24 h after electroporation. Recombinant Sindbis viral stocks are stored at –80 °C. We typically obtain viral stock with titer ranging 10^7–10^8 transducing units/ml which is enough for slice transduction or for in vivo injection.

3.2. Transducing Acute Neocortical Slices

All experiments are carried out in accordance with the guidelines published in the European Communities Council Directive of 24 November 1986 (86/609/EEC). Young Wistar rats (11–15 postnatal days old) are decapitated, the brains are quickly removed, and 300-µm-thick parasagittal sections of cerebral cortex are prepared as described (39) using a vibratome. Slices are incubated at room temperature for 30 min in ACSF containing (in mM) : 126 NaCl, 2.5 KCl, 1.25 NaH_2PO_4, 2 $CaCl_2$, 1 $MgCl_2$, 26 $NaHCO_3$, 20 glucose, 5 sodium pyruvate, bubbled with a mixture of 95 % O_2/5 % CO_2, and supplemented with 1 mM kynurenic acid. Then, slices are

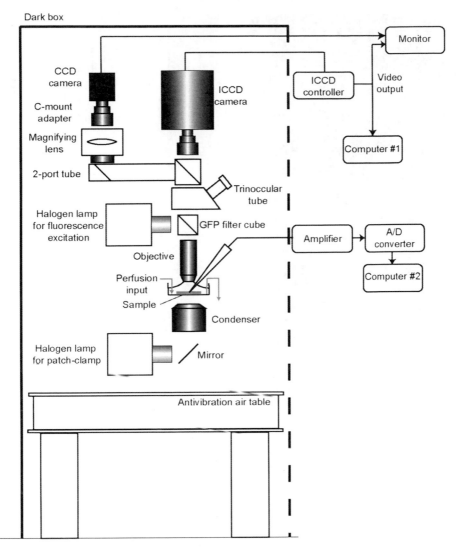

Fig. 1. Schematic representation of a setup for combined patch-clamp and bioluminescence recordings. Prior to imaging, samples are visualized by GFP epifluorescence using GFP filter set. Filters reside in a filter wheel (not shown), thus they can be removed from the light path during bioluminescence acquisition. Patch-clamp is performed under transillumination with a halogen lamp equipped with a high numerical aperture condenser. A front door (*dashed line*) in the *dark box* provides access to the setup.

transferred onto a millicell-CM membrane preequilibrated with culture medium (50 % MEM, 50 % HBSS, 6.5 g/l glucose, 10 U/ml penicillin, 10 µg/ml streptomycin). Slices are transduced with Sindbis pseudovirions by adding 5 µl of viral solution onto the slice and incubated at 35 °C in a standard tissue culture incubator with an humidified 5 % CO_2 atmosphere. 30 min after virus application, coelenterazine (final concentration: 10 µM) is added to the culture medium and slices are put back in the incubator and left overnight until the recording. The next day, slices are transferred in bubbled ACSF, incubated for at least 1 h at room temperature in the dark,

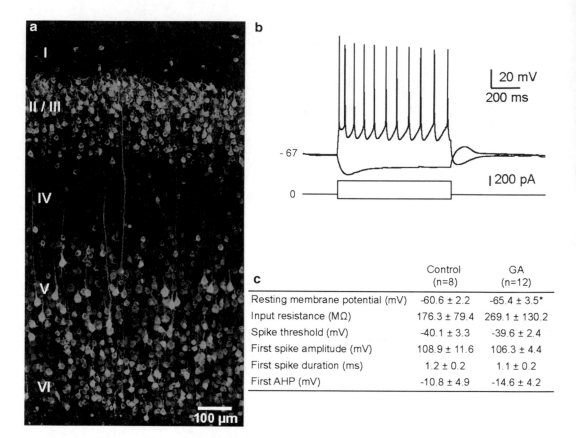

Fig. 2. Expression of bioluminescent sensors in neocortical slices. (a) Confocal reconstruction of a neocortical slice expressing the GA sensor following viral transduction. Neocortical layers are indicated. Cells expressing GA identified by their GFP fluorescence (*green*) are preferentially localized in layers II/III and V. A GA-expressing layer V pyramidal neuron has been patched and filled with biocytin for histochemical revelation (*red*). Note the characteristic pyramidal shape of the soma, the prominent apical dendrite ramifying in layers II–III and extending into layer I, and the axon projecting toward the white matter. (b) Electrophysiological responses of the same GA-expressing layer V pyramidal neuron to injection of current steps of −100 and 300 pA. (c) Electrophysiological properties (Mean ± SD) of control and GA-Expressing Layer V pyramidal neurons. *Statistically significant differences from control group ($P < 0.05$, Mann–Whitney test).

and then transferred into the recording chamber and perfused at 1–2 ml/min with ACSF. Recordings are typically performed 15–24 h after viral transduction. The incubation step in ACSF is necessary to equilibrate the slice with the recording solution and to avoid drift along the z-axis during recording.

GFP-photoprotein expression is essentially observed in cells displaying neuronal morphology consistently with the neurotropism of Sindbis virus. A majority of these neurons are located in cortical layers II/III and V and exhibit the typical morphology of pyramidal cells with a prominent apical dendrite extending from a pyramidal-shaped soma towards the apical surface (Fig. 2). As expected by the absence of subcellular targeting signals, GFP fluorescence is distributed in the soma and the whole dendritic tree, and also in the nucleus.

To test basic electrophysiological properties of transduced neurons, we use whole-cell patch-clamp recording on layer V pyramidal neurons expressing GA. Patch pipettes (3–6 MΩ) pulled from borosilicate glass (Hilgenberg GmbH, Malsfeld, Germany) are filled with 15 μl internal solution containing (in mM): 123 K-gluconate, 21 KCl, 10 HEPES, 3 MgCl$_2$, 0.2 EGTA, 4 NaATP, 0.4 NaGTP (pH = 7.2, 285/295 mOsm). Whole-cell recordings from GA expressing neurons, selected under epifluorescence illumination, are performed at room temperature with a conventional patch-clamp amplifier. Here we use an Axopatch1D (Molecular Devices, Sunnyvale, CA, USA) connected to a Digidata 1440 interface board (Molecular Devices). We evaluate several passive and active parameter as described previously (39): (1) input resistance, determined from the application of a –50 pA hyperpolarizing 800 ms current step, (2) action potential (AP) threshold, (3) amplitude of the first AP, measured from the threshold to the peak of the AP, (4) duration of the first AP measured at half amplitude, and (5) amplitude of the first afterhyperpolarization (AHP), measured from the spike threshold to the peak of the AHP.

Electrophysiological responses of GA expressing cells are compared to those of control layer V pyramidal neurons recorded from freshly prepared cortical slices. Overall, our results show that gross morphological and electrophysiological properties of layer V pyramidal neurons are essentially preserved after Sindbis transduction, overnight incubation and GA expression as previously reported in the cortex and other brain structures (40–43). Indeed, the firing frequency as a function of depolarizing current step intensity is similar between the GA-expressing and control groups. Values of electrophysiological parameters of GA expressing cells compared to those of control layer V neurons are also similar except for the resting membrane potential of GA expressing neurons which is slightly lower than the one of control neurons. Furthermore, values obtained in each group are in the range of those earlier reported for layer V pyramidal neurons (44, 45).

3.3. Estimation of the Intracellular Concentration of Indicator

Aequorin exhibits a low buffering capacity and alters minimally intracellular Ca²⁺ concentration in comparison with organic Ca²⁺ sensitive fluorescent dye (46). However, it is possible to estimate the concentration of any fluorescent probe following gene transfer. The procedure describes here is detailed for use of GFP based protein but it can be easily extended to other protein tagged with other fluorescent proteins.

The intracellular concentration of GA in layer V pyramidal neurons is determined using two-photon laser-scanning microscopy of GFP fluorescence, from comparison with a fluorescein calibration curve. The intensity of cytosolic GFP fluorescence is measured using a custom-built two-photon laser scanning microscope, equipped with a water immersion 60× objective (N.A. = 0.9;

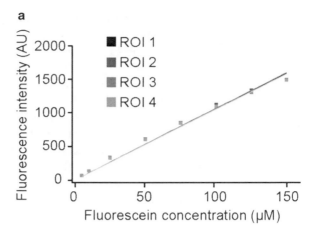

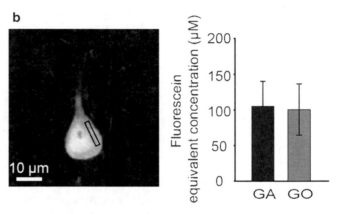

Fig. 3. Estimation of intracellular concentrations of GA and GO expressed in layer V pyramidal neurons with two-photon microscopy. (a) A fluorescence calibration curve is obtained with fluorescein solutions of known concentrations at pH 9. *Squares* represent experimental values measured from four different ROIs, which were fitted with linear functions (*solid lines*). Note that experimental values were almost superimposed, as well as linear fits. (b) *Left*: Single plane image of a GO-expressing layer V pyramidal neuron, where fluorescence intensity was measured at the cytoplasmic ROI delineated. *Right*: GFP-photoprotein cytosolic concentrations measured in GA- and GO-expressing neurons ($n = 21$ and 26 cells respectively). Data are expressed as mean ± standard deviation.

Olympus, Tokyo, Japan) and a Ti:sapphire laser (MaiTai HP; Spectra Physics, Mountain View, CA, USA) tuned to 920 nm. Fluorescence is detected with a photomultiplier tube (H9305, Hamamatsu Photonics, Massy, France). Image acquisition is controlled by MPScope software (47). Three optical plane sections separated by 2 μm are acquired in single layer V pyramidal neurons somata. For each cell, fluorescence intensity from a region of interest (ROI, Fig. 3) located in the cytoplasm is determined and the mean ROI fluorescence is calculated by averaging the ROI signals from the three optical sections. The same microscope settings are used to obtain a fluorescence calibration curve by imaging 100 μl droplets of fluorescein solutions (5–150 μM) at pH 9. The quantum yield and two-photon absorption cross section of enhanced-GFP (EGFP) in solution has been reported (48–50). Assuming that GFP moieties of the sensors exhibit in the cytosol the same optical characteristics as those of EGFP in solution, we calculated the concentration of the sensors using the following equation:

$$[GFP] = [Fluorescein] \times \frac{\sigma_{TPA\ fluorescein} \times \Phi_{fluorescein}}{\sigma_{TPA\ EGFP} \times \Phi_{EGFP}},$$

where σ_{TPA} is the two-photon absorption (TPA) cross section for fluorescein (26 GM; (48)) and EGFP (68 GM; (49)) and Φ the quantum yield for fluorescein (0.9; (48)) and EGFP (0.6; (50)).

Fluorescence values obtained for GA are equivalent to ~100 μM of fluorescein which corresponds to ~60 μM of GA in the transduced neurons (Fig. 3). Comparable values in the range of 10–60 μM fluorescein are reported for fluorescent genetically encoded Ca²⁺ probes expressed in pyramidal neurons of organotypic slice cultures following adenoassociated viral transduction (51). Expression levels by mean of viral vector are higher than those obtained in transgenic animals (51).

3.4. Imaging

In our experiments, coelenterazine is added directly into the culture medium of acute slice allowing it to diffuse freely across the porous membrane of the organotypic insert and then through the cell membrane. Once inside the cells, it spontaneously binds to the apoaequorin moiety reconstituting the active probe. We initially incubated slices with coelenterazine for 3 h in the incubator just before the recordings. However, we now obtain better results with overnight incubation while the probe is expressed by the cell after viral transduction.

For low light acquisition, we use the Photek intensified CCD camera ICCD225 in the mode "binning slice." This camera contains a photocathode connected to an electron-amplifier microchannel plate as intensifier. The same camera is used for fluorescence imaging using the "brightfield mode" with low gain. Special caution is required as the photocathode and the microchannel plate are at risk for burn-in when excessive light is applied. These kinds of cameras are equipped with a standard video interface (RS-170 or CCIR) making them easy to use with video peripherals such as TV monitor, video recorders, image processors, and image processing software. Furthermore, the frame rate (25–50 Hz for CCIR) is fast enough for most real-time applications and focusing.

For evaluating bioluminescence response to intracellular Ca²⁺ increase, cells are subjected to an electrophysiological stimulation sequence consisting in a series of 800 ms long current steps, ranging from −100 to 700 pA with 50 pA increment while bioluminescence is recorded using the "binary slice" photon counting mode at a video frame rate of 25 images per second. An example of a typical experiment is illustrated in Fig. 4. In this neuron, depolarizing

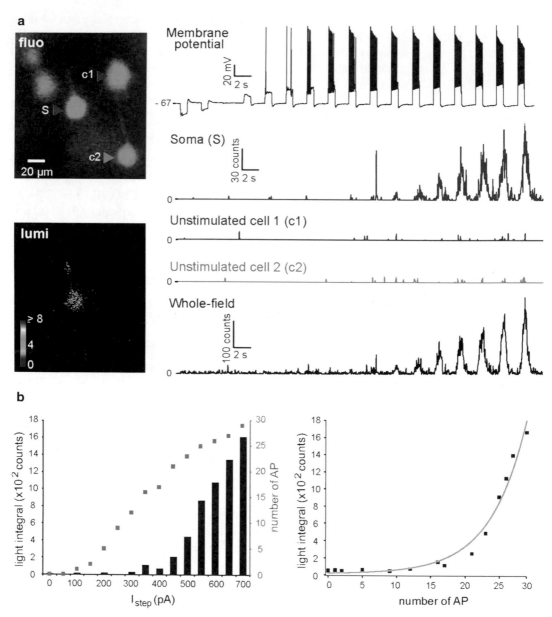

Fig. 4. Depolarization-evoked electrophysiological and light responses of a GO-expressing layer V pyramidal neuron. (**a**) *Upper left*: GFP fluorescence image showing the patched neuron stimulated with a sequence of current steps (*red arrowhead*) and two neighboring unstimulated neurons (*purple and blue arrowheads*, c1 and c2). *Bottom left*: The stimulated neuron is visible on the image of the luminescence emitted over the whole stimulation sequence. Pseudocolor scale indicates number of counts/pixel. *Right*: Electrophysiological (*black trace*) and luminescence (*red trace*) responses of the patched neuron to the stimulation sequence as well as the luminescence emitted by unstimulated neurons (*purple and blue traces*, c1 and c2). *Colored traces* represent the luminescent signal of the soma, and *gray trace* corresponds to the whole-field luminescence. Note the low background luminescence and the minimal light signal detected from unstimulated neurons. (**b**) *Left*: Luminescence response integrals (*black bars*) and number of action potentials emitted (*gray squares*) as a function of current step intensity (same neuron as in **a**). *Right*: Plot of the light integral as a function of action potentials in the same neuron.

steps of increasing amplitude elicit AP firing of increasing frequency. Prior to a threshold depolarizing step, the recorded light signal is close to zero, consistent with the low background luminescence of photoproteins (15, 16, 19). From the sixth depolarizing step (300 pA), bioluminescence is consistently emitted from the stimulated neuron in response to each step. Light emission shows a marked increase with increasing stimulation intensity. Light responses to each depolarizing step increase during the step to reach a maximum at the end of the step. Responses rapidly declined after the stimulation. Only minimal light signals is detected from neighboring unstimulated neurons during the stimulation sequence, indicating that Ca^{2+}-activated bioluminescence occurs primarily in the single stimulated neuron (Fig. 4). This is assessed by the bioluminescence image integrated over the whole stimulation sequence, where the soma and proximal dendrite of the stimulated cell are visible. The bioluminescence response of this neuron is clearly detectable on the whole field recording. Only 30% of the total bioluminescence emitted during the stimulation sequence comes from the soma of the stimulated neuron, suggesting that a large part of the whole field bioluminescence response derives from the dendritic arbor of the stimulated neuron.

The absence of coenlenterazine after the initial loading prevents the reconstitution of aequorin following light emission, eventually leading to an exhaustion of active photoprotein. In order to estimate the fraction of aequorin consumed during a given experiment, it is necessary to consume all available active aequorin using a protocol that achieves saturating intracellular Ca^{2+} elevation. While in cell culture, this can be performed using Ca^{2+} ionophore such as ionomycin, in acute slice, the poor diffusion of ionomycin makes this procedure ineffective. Thus, we elicit large intracellular Ca^{2+} increases at the end of the recording by applying successive 30 s voltage-clamp steps at 0 mV, thereby activating voltage-gated Ca^{2+} channels, until the bioluminescence response is exhausted. This is achieved typically within 3–8 steps. The total light emitted from a cell is thus calculated as the light integral over the whole experiment and is called L_{tot} (Fig. 5).

Fractional light response is then calculated as the ratio of the integral of the light emitted during single stimulation sequences (L_{seq}) over the light integral. Figure 5 shows an example of the light and electrophysiological responses of a GA-expressing neuron to the stimulation sequence, followed by the light signal recorded during the voltage-clamp protocol. Consumption of the photoprotein with voltage clamp steps is evidenced by the decline of the light response, which is almost exhausted at the fifth step. Essentially similar results are obtained with GO. Linear regression of the plot of L_{seq} against L_{tot} for all tested cells has a slope of 0.01 (Fig. 4), indicating that the intensity of light responses to stimulation sequences is independent of GFP-photoprotein sensors expression

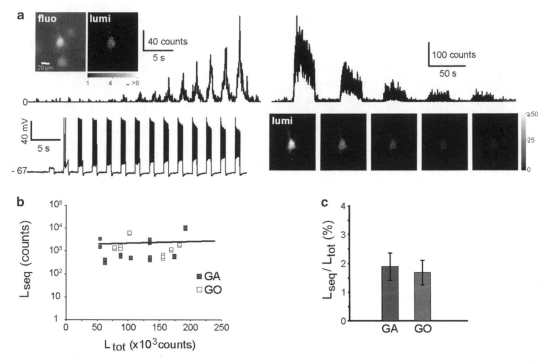

Fig. 5. Fractional light responses in layer V pyramidal neurons. (**a**) *Left*: Light and electrophysiological responses of a GA-expressing neuron to the current-clamp stimulation sequence. The recorded neuron (center of the fluorescence image) is visible on the image of the luminescence emitted over the whole stimulation sequence. *Right*: Light responses of the same neuron to five successive 30-s voltage-clamp steps at 0 mV, performed at the end of the experiment. Each image corresponds to the light emitted during one step. Note the step-to-step decrease of the light response, suggesting a massive consumption of the photoprotein. (**b**) Light emitted during single stimulation sequences (L_{seq}) expressed as a function of total light emitted during the experiment (L_{tot}) in the GA-and GO-expressing cell groups. Linear regression of this plot had a slope of 0.01 (*gray line*). (**c**) Histogram of mean fractional light responses in the GA- and GO-expressing groups.

levels. This suggests that sensors concentrations achieved using our viral expression method (see below for the determination of sensor concentration) is superior to intracellular Ca^{2+} concentrations reached during the stimulation sequence.

3.5. Analysis of Multi-unit Bioluminescence Recordings

Unlike with fluorescent genetically encoded Ca^{2+} sensors, imaging of neuronal ensembles using bioluminescence reporters does not require optical sectioning (typically acquired at 50–10 full frame/ second), which results in a significant loss of information. Additionally while the low level of light emission is overcome by the use of intensified CCD camera, the low number of emitted photons allows a compact storage of data. Indeed, the IFS32 Photek software compresses data using a run-length encoding (RLE) strategy that keeps only the *x–y* coordinate of a photon hit on the CCD sensor and the number of adjacent hits on the same pixel row. Consequently, bioluminescence recordings generate files several thousand fold smaller than fluorescence imaging.

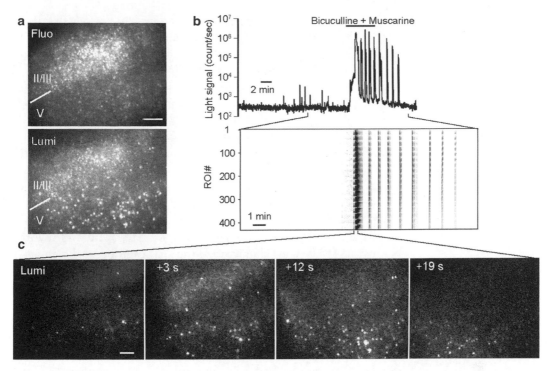

Fig. 6. Multi-unit bioluminescence recordings of GA-expressing cortical slices. (**a**) *Top*: Wide field fluorescence picture of a cortical slices taken with the intensified CCD camera 16 h after viral transduction. *Bottom*: Image of the light integrated over the response to the bath application of a mixture bicuculine (10 μM) and muscarine (10 μM). Note that several cell bodies are clearly visible. (**b**) *Top*: light responses of the imaging field to the bicuculine/muscarine application. *Bottom*: *gray scale rasterplot* of the light response of 423 square ROIs (60 × 60 μm). ROIs are arranged according to an 18 × 24 array covering the whole imaging field. ROI #1 and #432 are located at the upper left and lower right corner respectively. (**c**) Snapshots of the bioluminescence (integration time: 1 s) taken at different stage during the long-lasting light response. Scale bars: 200 μm.

To analyze Ca^{2+} activity of multiple GA-expressing neurons in cortical slices, we import data files generated by the program IFS32 controlling the intensified camera into the software Visilog 6.3 (Noesis, France) and data recorded during the period of interest are converted into a movie using a custom written program. Then, the imaging field is subdivided into an 18 × 24 arrays of 60 × 60 μm ROI and light signal from each ROI is read and plotted along an *x*-axis representing time. An example of this procedure is illustrated in Fig. 6 where we examined with a 10× objective (imaging filed: 1.4 mm × 1 mm) at video frame rate (25 full frame/second; integration time: 40 ms) the stereotypical responses of the cortical network to the combined application of the GABAA receptor antagonist bicuculline (10 μM) and the selective cholinergic agonist muscarine (10 μM) on network activity. The picture of the integrated light responses to the application of the drugs reveals hot spots of light emission matching the location of GFP fluorescent neurons (Fig. 6a). It also suggests that light is emitted by neurons from all cortical layers. Signal integrated on the whole imaging field show that bicuculline and muscarine cause a massive long-lasting (~30 s) increase in neuronal activity which is followed by oscillations constituted of brief burst (Fig. 6b). The rasterplot analysis using the procedure described above confirms that light responses, both the long-lasting one and the brief bursts, involved a large portion of the neurons of the imaging field. It also shows that the spatiotemporal course of the two responses is different. While the spatial distribution of long-lasting response evolves over time, the different brief bursts of activity appear very similar between each other (Fig. 6b). This is evidenced when we examined successive snapshots within the long-lasting light response, which show that this neuronal activity spreads from one side of the imaging field and crosses the entire field. At that stage, light signals of individual ROI are exported into Matlab (The Mathworks, Natick, MA, USA) to characterize the spatiotemporal dynamics of these activities. Several kind of analysis can thus be performed like calculating the cross-correlogram between ROIs or determining the movement of the center of gravity of the light response over time. These two analyses would give respectively the front speed and the overall mean speed of the response.

These results show that the study of the spatiotemporal dynamics of neuronal complex ensembles can be achieved at both single cell and whole field levels with Ca^{2+} bioluminescence recordings using photoproteins such as aequorin and obelin.

References

1. Shimomura O, Johnson FH, Saiga Y (1962) Extraction, purification and properties of aequorin, a bioluminescent protein from the luminous hydromedusann Aequorea. J Cell Comp Physiol 59:223–239

2. Ridgway EB, Ashley CC (1967) Calcium transients in single muscle fibers. Biochem Biophys Res Commun 29:229–234

3. Stinnakre J, Tauc L, Saito N (1972) Demonstration with aequorine photoprotein of variations of calcium activity in Aplysia neurons. J Physiol Paris 65(Suppl):308A

4. Llinas R, Blinks JR, Nicholson C (1972) Calcium transient in presynaptic terminal of squid giant synapse: detection with aequorin. Science 176:1127–1129

5. Moreau M, Guerrier P, Doree M, Ashley CC (1978) Hormone-induced release of intracellular Ca^{2+} triggers meiosis in starfish oocytes. Nature 272:251–253

6. Pinton P, Rimessi A, Romagnoli A, Prandini A, Rizzuto R (2007) Biosensors for the detection of calcium and pH. Methods Cell Biol 80:297–325

7. Shimomura O, Johnson FH (1975) Regeneration of the photoprotein aequorin. Nature 256:236–238

8. Wilson T, Hastings JW (1998) Bioluminescence. Annu Rev Cell Dev Biol 14:197–230

9. Vysotski ES, Lee J (2004) Ca^{2+}-regulated photoproteins: structural insight into the bioluminescence mechanism. Acc Chem Res 37:405–415

10. Hastings JW, Mitchell G, Mattingly PH, Blinks JR, Van Leeuwen M (1969) Response of aequorin bioluminescence to rapid changes in calcium concentration. Nature 222:1047–1050

11. Shimomura O, Johnson FH (1970) Calcium binding, quantum yield, and emitting molecule in aequorin bioluminescence. Nature 227:1356–1357

12. Tricoire L, Tsuzuki K, Courjean O, Gibelin N, Bourout G, Rossier J, Lambolez B (2006) Calcium dependence of aequorin bioluminescence dissected by random mutagenesis. Proc Natl Acad Sci USA 103:9500–9505

13. Tsuji FI, Ohmiya Y, Fagan TF, Toh H, Inouye S (1995) Molecular evolution of the Ca(2+)-

binding photoproteins of the Hydrozoa. Photochem Photobiol 62:657–661

14. Morin JG, Hastings JW (1971) Biochemistry of the bioluminescence of colonial hydroids and other coelenterates. J Cell Physiol 77: 305–312

15. Illarionov BA, Frank LA, Illarionova VA, Bondar VS, Vysotski ES, Blinks JR (2000) Recombinant obelin: cloning and expression of cDNA purification, and characterization as a calcium indicator. Methods Enzymol 305: 223–249

16. Markova SV, Vysotski ES, Blinks JR, Burakova LP, Wang BC, Lee J (2002) Obelin from the bioluminescent marine hydroid Obelia geniculata: cloning, expression, and comparison of some properties with those of other Ca²⁺-regulated photoproteins. Biochemistry 41: 2227–2236

17. Drobac E, Tricoire L, Chaffotte AF, Guiot E, Lambolez B (2010) Calcium imaging in single neurons from brain slices using bioluminescent reporters. J Neurosci Res 88:695–711

18. Moreau M, Vilain JP, Guerrier P (1980) Free calcium changes associated with hormone action in amphibian oocytes. Dev Biol 78: 201–214

19. Allen DG, Blinks JR, Prendergast FG (1977) Aequorin luminescence: relation of light emission to calcium concentration—a calcium-independent component. Science 195:996–998

20. Kendall JM, Sala-Newby G, Ghalaut V, Dormer RL, Campbell AK (1992) Engineering the CA(2+)-activated photoprotein aequorin with reduced affinity for calcium. Biochem Biophys Res Commun 187:1091–1097

21. Tsuzuki K, Tricoire L, Courjean O, Gibelin N, Rossier J, Lambolez B (2005) Thermostable mutants of the photoprotein aequorin obtained by in vitro evolution. J Biol Chem 280: 34324–34331

22. Shimomura O, Musicki B, Kishi Y (1989) Semi-synthetic aequorins with improved sensitivity to Ca²⁺ ions. Biochem J 261:913–920

23. Llinas R, Sugimori M, Silver RB (1992) Microdomains of high calcium concentration in a presynaptic terminal. Science 256:677–679

24. Llinas R, Sugimori M, Silver RB (1995) The concept of calcium concentration microdomains in synaptic transmission. Neuropharmacology 34:1443–1451

25. Robert V, De Giorgi F, Massimino ML, Cantini M, Pozzan T (1998) Direct monitoring of the calcium concentration in the sarcoplasmic and endoplasmic reticulum of skeletal muscle myotubes. J Biol Chem 273:30372–30378

26. Baubet V, Le Mouellic H, Campbell AK, Lucas-Meunier E, Fossier P, Brulet P (2000) Chimeric green fluorescent protein-aequorin as bioluminescent Ca²⁺ reporters at the single-cell level. Proc Natl Acad Sci USA 97:7260–7265

27. Martin JR, Rogers KL, Chagneau C, Brulet P (2007) In vivo bioluminescence imaging of Ca signalling in the brain of Drosophila. PLoS One 2:e275

28. Rogers KL, Picaud S, Roncali E, Boisgard R, Colasante C, Stinnakre J, Tavitian B, Brulet P (2007) Non-invasive in vivo imaging of calcium signaling in mice. PLoS One 2:e974

29. Rogers KL, Stinnakre J, Agulhon C, Jublot D, Shorte SL, Kremer EJ, Brulet P (2005) Visualization of local Ca²⁺ dynamics with genetically encoded bioluminescent reporters. Eur J Neurosci 21:597–610

30. Naumann EA, Kampff AR, Prober DA, Schier AF, Engert F (2010) Monitoring neural activity with bioluminescence during natural behavior. Nat Neurosci 13:513–520

31. Badminton MN, Sala-Newby GB, Kendall JM, Campbell AK (1995) Differences in stability of recombinant apoaequorin within subcellular compartments. Biochem Biophys Res Commun 217:950–957

32. Gorokhovatsky AY, Marchenkov VV, Rudenko NV, Ivashina TV, Ksenzenko VN, Burkhardt N, Semisotnov GV, Vinokurov LM, Alakhov YB (2004) Fusion of Aequorea victoria GFP and aequorin provides their Ca(2+)-induced interaction that results in red shift of GFP absorption and efficient bioluminescence energy transfer. Biochem Biophys Res Commun 320:703–711

33. Morin JG, Hastings JW (1971) Energy transfer in a bioluminescent system. J Cell Physiol 77:313–318

34. Pologruto TA, Yasuda R, Svoboda K (2004) Monitoring neural activity and (Ca²⁺) with genetically encoded Ca²⁺ indicators. J Neurosci 24:9572–9579

35. Ehrengruber MU (2002) Alphaviral gene transfer in neurobiology. Brain Res Bull 59: 13–22

36. Altman-Hamamdzic S, Groseclose C, Ma JX, Hamamdzic D, Vrindavanam NS, Middaugh LD, Parratto NP, Sallee FR (1997) Expression of beta-galactosidase in mouse brain: utilization of a novel nonreplicative Sindbis virus vector as a neuronal gene delivery system. Gene Ther 4:815–822

37. Gwag BJ, Kim EY, Ryu BR, Won SJ, Ko HW, Oh YJ, Cho YG, Ha SJ, Sung YC (1998) A neuron-specific gene transfer by a recombinant

defective Sindbis virus. Brain Res Mol Brain Res 63:53–61

38. Bredenbeek PJ, Frolov I, Rice CM, Schlesinger S (1993) Sindbis virus expression vectors: packaging of RNA replicons by using defective helper RNAs. J Virol 67:6439–6446

39. Cauli B, Audinat E, Lambolez B, Angulo MC, Ropert N, Tsuzuki K, Hestrin S, Rossier J (1997) Molecular and physiological diversity of cortical nonpyramidal cells. J Neurosci 17: 3894–3906

40. Hepp R, Tricoire L, Hu E, Gervasi N, Paupardin-Tritsch D, Lambolez B, Vincent P (2007) Phosphodiesterase type 2 and the homeostasis of cyclic GMP in living thalamic neurons. J Neurochem 102:1875–1886

41. D'Apuzzo M, Mandolesi G, Reis G, Schuman EM (2001) Abundant GFP expression and LTP in hippocampal acute slices by in vivo injection of sindbis virus. J Neurophysiol 86: 1037–1042

42. Chen BE, Lendvai B, Nimchinsky EA, Burbach B, Fox K, Svoboda K (2000) Imaging high-resolution structure of GFP-expressing neurons in neocortex in vivo. Learn Mem 7: 433–441

43. Diaz LM, Maiya R, Sullivan MA, Han Y, Walton HA, Boehm SL 2nd, Bergeson SE, Mayfield RD, Morrisett RA (2004) Sindbis viral-mediated expression of eGFP-dopamine D1 receptors in situ with real-time two-photon microscopic detection. J Neurosci Methods 139:25–31

44. Andjelic S, Gallopin T, Cauli B, Hill EL, Roux L, Badr S, Hu E, Tamas G, Lambolez B (2009) Glutamatergic nonpyramidal neurons from neocortical layer VI and their comparison with pyramidal and spiny stellate neurons. J Neurophysiol 101:641–654

45. Christophe E, Doerflinger N, Lavery DJ, Molnar Z, Charpak S, Audinat E (2005) Two populations of layer v pyramidal cells of the mouse neocortex: development and sensitivity to anesthetics. J Neurophysiol 94:3357–3367

46. Brini M, Marsault R, Bastianutto C, Alvarez J, Pozzan T, Rizzuto R (1995) Transfected aequorin in the measurement of cytosolic Ca^{2+} concentration ((Ca^{2+})c). A critical evaluation. J Biol Chem 270:9896–9903

47. Nguyen QT, Tsai PS, Kleinfeld D (2006) MPScope: a versatile software suite for multiphoton microscopy. J Neurosci Methods 156:351–359

48. Albota MA, Xu C, Webb WW (1998) Two-photon fluorescence excitation cross sections of biomolecular probes from 690 to 960 nm. Appl Opt 37:7352–7356

49. Blab GA, Lommerse PHM, Cognet L, Harms GS, Schmidt T (2001) Two-photon excitation action cross-sections of the autofluorescent proteins. Chem Phys Lett 350:71–77

50. Patterson GH, Knobel SM, Sharif WD, Kain SR, Piston DW (1997) Use of the green fluorescent protein and its mutants in quantitative fluorescence microscopy. Biophys J 73: 2782–2790

51. Wallace DJ, Meyer zum Alten Borgloh S, Astori S, Yang Y, Bausen M, Kugler S, Palmer AE, Tsien RY, Sprengel R, Kerr JN, Denk W, Hasan MT (2008) Single-spike detection in vitro and in vivo with a genetic Ca^{2+} sensor. Nat Methods 5:797–804

Chapter 3

Calcium Imaging of Neural Activity in the Olfactory System of Drosophila

Antonia Strutz, Thomas Völler, Thomas Riemensperger, André Fiala, and Silke Sachse

Abstract

Many animals are able to detect a plethora of diverse odorants using arrays of odorant receptors located on the olfactory organs. The olfactory information is subsequently encoded and processed by an overlapping, combinatorial activity of neurons forming complex neural circuits in the brain. In order to functionally dissect this neural circuitry, optical recording techniques allow visualizing spatial as well as temporal aspects of odor representations in populations of olfactory neurons. The fruit fly *Drosophila melanogaster* has emerged as a highly suitable model system for olfactory research as it allows for the combination of genetic, molecular and physiological analyses. Genes of interest can be ectopically expressed in target regions using different binary transcriptional systems. Thereby, fluorescent calcium indicators can be expressed to monitor neuronal activity in genetically defined subsets of neurons. In this chapter we describe various available genetically encoded calcium sensors (GECIs) and the binary transcriptional systems available for *Drosophila* to express these GECIs in olfactory neurons. We will explain step-by-step methods for fly brain preparation, introduce different odor application devices, and describe the components needed using a widefield or two-photon imaging system including data acquisition and analysis. Overall, this review provides a guideline for optically monitoring the spatiotemporal neuronal activity evoked by odorants in the *Drosophila* brain.

Key words: *Drosophila melanogaster*, Genetically encoded calcium indicators, Binary transcriptional systems, Cameleon, G-CaMP, Olfaction, Optical recording, Two-photon imaging, Antennal lobe, Mushroom body, Insect brain

1. Background and Overview

The olfactory system of the fruit fly *Drosophila melanogaster* represents a favorable model system for the analysis and dissection of a complex neuronal circuitry (1, 2). Initially, *Drosophila* has been used as a genetic model organism, and studies on olfaction in fruit flies have focused on the identification of genes involved in odor-guided behavior (3–6). In addition, neuroanatomical studies have

Jean-René Martin (ed.), *Genetically Encoded Functional Indicators*, Neuromethods, vol. 72,
DOI 10.1007/978-1-62703-014-4_3, © Springer Science+Business Media, LLC 2012

clarified the structure of the fly's odor-sensing organs and the connectivity of olfactory neurons in the brain to a fair degree (7, 8). Physiological studies, however, have for a long time remained difficult in fruit flies due to the small size of neurons, and electrophysiological recordings have mostly been performed in larger insects, e.g., locusts, cockroaches, moths, or honeybees. In recent years electrophysiological recordings from olfactory neurons in *Drosophila* not only at the level of peripheral sensory cells (9–12) but also from central brain neurons (13–17) have helped to deepen understanding of how odors are encoded at the level of individual cells. However, the advent of genetically encoded fluorescent indicators to monitor neuronal activity has opened a route to combine the genetic advantages of the fruit fly with optical methods to monitor neuronal activity across larger populations of neurons (18). Here, we describe how genetically encoded calcium sensors (GECIs) can be used to monitor spatiotemporal activity in neuronal subpopulations of the *Drosophila* olfactory system.

1.1. The Olfactory System of Drosophila melanogaster

Adult fruit flies detect odors with ~1,200 olfactory sensory neurons (OSNs) located at the third antennal segments and the maxillary palps (1, 2). Up to four OSNs are integrated into sensillae of various morphological types (10), and extend ciliary protrusions into fluid filled, perforated, hair-like cuticular structures (19, 20). Here, odorant receptors (ORs) located on the surface of the ciliary membranes bind volatile substances as odorants. The majority (~2/3) of OSNs express one, in some instances two specific ORs each out of 62 OR transcripts present in adult flies (21–25). These "classical OR genes" are coexpressed with the more ubiquitous receptor Or83b (26), also called "olfactory receptor coreceptor (Orco)" that forms heterodimeric proteins with the ligand-specific ORs (27, 28). Recently a novel class of ionotropic olfactory receptors (IRs) has been discovered (29, 30) that might account for the remaining ~1/3 OSNs. For both types of olfactory receptors, ORs and IRs, signal transduction has been shown to be mediated via ionotropic channel proteins (28, 31), but for ORs in addition a metabotropic signaling pathway via cyclic nucleotides is also involved (31). In summary, a specific response profile of each OSN with respect to odor-specificity is, therefore, determined by the expression of the particular OR (12) or IR (29). However, OSNs do not only show specificity with respect to the expression of a certain receptor, but also with respect to the brain region targeted by their axonal terminals. OSNs innervate the antennal lobes (AL), the primary olfactory neuropil of insects. The logic of targeted projection is straightforward: those OSNs that express a particular OR arborize in the same one or very few glomeruli of the AL (32, 33). As each OSN responds with a certain degree of ligand-specificity odors are represented at the level of the AL in terms of spatiotemporal, glomerular activity patterns (34, 35).

Afferent olfactory information is conveyed from the ALs to "higher brain centers" via olfactory projection neurons (PNs). In most cases PNs arborize in the AL within single glomeruli, but multiglomerular PNs also exist (36, 37). However, the AL is not merely a simple input–output relay. On the one hand, PNs form also presynaptic termini with the ALs' glomeruli, providing already a step of odor information processing. On the other hand, local interneurons (LNs) interconnect the ~49 glomeruli of the AL. The anatomical structure of LNs is diverse with respect to their arborization patterns, with some LNs innervating only few glomeruli, whereas others target many glomeruli more globally (17). In addition, the transmitter substances released by LNs are also not homogeneous. In most cases LNs are inhibitory and release GABA (17), but in some cases LNs exert excitatory transmission through acetylcholine (15, 38). Recently, it has been shown that excitatory LNs are electrically coupled through gap junctions, implicating a role of LNs for generally enhancing depolarizations evoked by faint concentrations of odors (39, 40). Inhibitory GABAergic LNs are considered to enhance decorrelation of odor-evoked PN responses, ultimately improving separability of distinct odors (13). In addition to LNs, the AL is a target for modulatory neurons releasing biogenic amines, e.g., serotonin (41) or octopamine (42), and for modulatory peptides (43). Overall, the AL circuitry performs a complex computational processing of olfactory information, and even learning-induced changes in odor representations have been detected in the AL (44).

About 150 PNs leave the AL and convey odor information to the mushroom body and the lateral horn. Whereas uniglomerular PNs arborize in the lateral horn and, en passant, in the main olfactory input region of the mushroom body, the calyx, multiglomerular PNs bypass the calyx and project directly to the lateral horn. Neuroanatomical studies have revealed that the arborizations of PNs are stereotypic with respect to their target areas (36, 37, 45, 46), but the lateral horn remains functionally not very well investigated as yet. However, the mushroom body has been a major subject of research since several decades due to its well substantiated role in associative olfactory learning (47–50). Here, the ~150 PNs synapse onto ~2,000–2,500 Kenyon cells (KCs), the intrinsic neurons of the mushroom body (51). The relatively high firing threshold and the crosswise interconnections between PNs and KCs speak in favor of an odor-coding principle called "sparse code", which means that only very few KCs are exclusively responsive to a distinct odor (52, 53), a principle that appears favorable for associating distinct odors with rewarding or punitive stimuli (48). Taken together, both at the level of the AL and at the level of the mushroom body odor information is processed, and the information about a particular odor is distributed across relatively large numbers of neurons. In order to analyze how odor information is processed

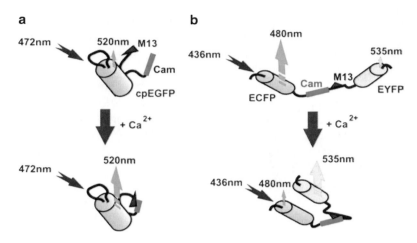

Fig. 1. Schematic illustration of a FRET-based sensor and a single-chromophore calcium sensor. (**a**) Using FRET-based, two-chromophore calcium sensors a Ca^{2+} influx can be detected by an increase in fluorescence emission of the acceptor (in the case of Cameleon EYFP) and a decrease in fluorescence emission of the donor (in the case of Cameleon ECFP). (**b**) Using single-chromophore calcium sensors, e.g., G-CaMP, Ca^{2+} dynamics are reflected in the intensity of the fluorescence emission of the circularly permuted GFP variant (cpEGFP).

at successive steps of integration it is of course helpful to observe the spatial distribution of neuronal activity at each relay station, a task for which optical calcium indicators can be used.

1.2. Genetically Encoded Calcium Indicators

The idea that odor representations are spatially represented at the level of the AL in terms of glomerular activity patterns has first been demonstrated in *Drosophila* by desoxyglucose mapping (54, 55). As conceptually important as these findings were, technological progress has advanced the field of research, and nowadays genetically encoded calcium indicators (GECIs) make it possible to monitor neuronal activity in subpopulations of neurons defined by a common genetic and functional identity, e.g., OSNs, PNs, and LNs. Since the invention of monitoring neuronal activity using calcium imaging in fruit flies (34) a variety of genetically encoded sensor proteins have been described by now. Most sensor proteins rely on variants of the green fluorescent protein (GFP; but see ref. (56) for an alternative method exploiting bioluminescence). For those GECIs based on fluorescence, two strategies to construct GECIs have been pursued, FRET-based sensors with two chromophores and circularly permuted GFP variants with one chromophore (Fig. 1). Both types of sensors rely on conformational changes induced by calcium binding, ultimately changing light emission of the fluorescence sensor. FRET-based calcium sensors consist of two GFP variants, a donor and an acceptor, which are fused to a calcium binding domain and a target peptide that mediates a conformational change upon calcium binding. If the donor chromophore, typically a light blue variant of GFP, is excited at a wavelength of ~440 nm, a cyan emission of 480 nm can be detected.

Table 1
Selected genetically encoded fluorescent sensor proteins

GECI	Type of sensor	Reference for sensor	Reference for fly strains
Cameleon 2.1	FRET-based	(68)	(34, 96)
Cameleon 3.6	FRET-based	(92)	(62)
Cameleon 6.1	FRET-based	(89)	(80)
TN-XL	FRET-based	(90)	(90)
TN-XXL	FRET-based	(93)	(93)
G-CaMP 1.3	Single-chromophore	(59)	(35)
G-CaMP 1.6	Single-chromophore	(91)	(97)
G-CaMP 3.0	Single-chromophore	(94)	(94)
Camgaroo	Single-chromophore	(98, 99)	(100)

The conformational change induced by Ca^{2+} now leads to an increase in fluorescence energy transfer (Förster resonance energy transfer, FRET) from the donor to the acceptor, typically a yellow variant of GFP. An increase in intracellular Ca^{2+} can be detected by a decrease in donor emission and an increase in acceptor emission (Fig. 1b). Since the first descriptions of FRET-based calcium sensors (57, 58) a variety of improved versions have been engineered with various fluorophores or calcium binding domains, e.g., from calmodulin interacting with a calmodulin target peptide in the case of the cameleon-type and G-CaMP-type GECIs, and troponin C in the case of troponin-based sensors. Many of these GECIs have been used also in *Drosophila* (Table 1).

An alternative idea to construct a GECI has been invented by Nakai et al. (59). Here, a GFP molecule has been modified so that novel N- and C-termini of the protein have been introduced to calcium binding domain and target peptide have been fused to (Fig. 1a). Ca^{2+} binding now changes the protonation state of the chromophore, ultimately causing an increase in emission intensity. Also here several types have been described (Table 1), and this type of GECI represents nowadays the most commonly used calcium sensor due to its easier handling as it requires detecting only one single emission wavelength.

Rigorous comparisons between a number of GECIs have been performed in transgenic mice in vivo (60), in brain slices (61) and in motoneurons of the larval neuromuscular junction of *Drosophila* (62, 63). These elaborate studies have provided insights into the properties of various sensors under optimized experimental conditions in which neuronal activation is controllable and optical access is unobstructed. In order to exploit these tools for the analysis of brain function it is often required to use GECIs in the central brain of largely intact animal, a situation that differs from brain slices or neuromuscular junction preparations. First, signal intensities evoked by physiological stimuli are often much smaller than electrophysiologically evoked signals. Second, small movements of the structures under investigation cannot be completely avoided. Third, optical access to neuronal structures under investigation is often affected by surrounding tissue and optical resolution is difficult due to the small size of fine and widespread arborizing structures. To decide which GECIs to use it is important to ask which parameters are critical for the quality of the optical recording. The first point to be considered is not the absolute change in fluorescence emission upon neuronal activation (signal) but the signal-to-noise ratio, which can be defined as the ratio of the signal change divided by the variance of the signal. Both properties of the particular GECI (e.g., the binding constant of the calcium binding domain to intracellular Ca^{2+} and the resulting kinetics of Ca^{2+} binding), the physiological properties of the measured cellular populations, the expression level of the GECI, potential movement of the preparation and the optical measurement method (e.g., widefield or two-photon microscopy) influence the signal-to-noise ratio. A critical parameter is the dissociation constant (K_D) of the GECI's binding sites for Ca^{2+} as it directly determines to a large degree the kinetics and dynamic range of the sensor. Different GECIs with different Ca^{2+} K_D values have been described, and thoroughly characterized by Reiff et al. (63) and Hendel et al. (62) in *Drosophila*. Whereas The GECI Yellow Cameleon 3.6 shows a relatively high Ca^{2+} affinity and fast kinetics, the GECI TN-XL is at the opposite side of the range (62). Which GECI serves best for a particular question depends, however, on the dynamic range of Ca^{2+} influx to be detected and whether transient neural activity with fast signal kinetics is intended to be monitored. In addition, the actual expression level of the GECI is also a critical factor determining signal kinetics. It should also be noted that for some applications a rather high baseline fluorescence of the GECI is convenient, in particular if for example Ca^{2+} signals through the pigmented cuticle are monitored (64, 65). We would like to mention that the recently described GECIs TN-XXL (93) and G-CaMP 3.0 (94) offer strong improvements in terms of signal kinetics and signal amplitude. Here, we do not aim at comparing different GECIs in detail, but rather describe technically how to employ these sensors in order to visualize odor-evoked neuronal Ca^{2+} activity in subpopulations of neurons within the *Drosophila* olfactory system.

2. Fly Strains, Materials, and Setup

2.1. Binary Transcriptional Systems for Selective GECI Expression

The model organism *Drosophila melanogaster* has the great advantage of a sequenced genome, which is modifiable through a large number of genetic tools. Due to their combinatorial applicability, binary transcriptional systems belong to the most valuable tools. The approach makes use of exogenous expression systems that are unrelated to DNA sequences of the *Drosophila* genome, and which operate under the control of the endogenous transcription machinery.

To achieve directed gene expression an exogenous transcription factor is inserted via P-element-mediated transformation relatively randomly into the fly's genome, which leads to its expression controlled by a promoter or enhancer sequence, respectively. Alternatively, the transcription factor can be inserted directly downstream of a specific promoter sequence. This fly strain ("driver line") carrying the transcription factor determines the cell-type-specific location of the expression. In a second transgenic fly strain ("effector line") the reporter, in this case the GECI sequence, is coupled to a cognate target sequence of the transcription factor, and this DNA construct is also inserted into the fly's genome. If both fly strains are crossed the endogenous expression control leads to a spatially and/or temporally restricted GECI expression depending on the promoter activity (indirectly) regulating the effector transcription. Accomplishing the genetic modifications in different fly strains allows the combination of many possible effectors and reporters (Table 2).

The first binary transcriptional system developed in the fly has been the *GAL4-UAS system* (66, 67). The transcription factor GAL4 from yeast initiates the transcription of several genes by interacting with the GAL4 Upstream Activating Sequence (UAS). For general applicability the system was subsequently adapted by Brand and Perrimon (67), which inter alia designed optimized UAS cassettes consisting of several tandemly arranged GAL4 binding sites. Nowadays a huge variety of GAL4 lines is available (e.g., at the Bloomington stock center, USA, or the *Drosophila* Genetic Resource Center in Kyoto, Japan) which can be used to drive the expression of a particular GECI in defined populations of neurons.

Using the GAL4-UAS-system GECIs have been expressed in various olfactory neurons along the olfactory pathway. The fluorescence calcium sensor Cameleon 2.1 (68) was one of the first GECIs expressed via the GAL4-UAS system (34) under control of the GH146-GAL4 that targets about 60 % of olfactory PNs. In 2003, Wang et al. (35) have used the GECI G-CaMP 1.3 (59) for functional two-photon imaging in *Drosophila*. In their seminal paper Wang and coworkers expressed G-CaMP 1.3 in single glomeruli (ORx-GAL4), in OSNs expressing the OR coreceptor ORCO (i.e., majority of OSNs) using Or83b-GAL4 (26) or in the majority

Table 2
Available binary transcriptional systems in *Drosophila*

Binary transcriptional system	Origin	Discovery/development	Repressor	Available effector fly strains	Available reporter fly strains
GAL4—UAS	*Saccharomyces cerevisiae*	(66, 67)	GAL80	Many (e.g., http://flystocks.bio.indiana.edu)	Many (e.g., http://flystocks.bio.indiana.edu/)
LexA—LexAop	*Escherichia coli*	(73, 74, 101)	GAL80 (for LexA::GAD) no independent repressor for LexA::VP16	LexA::VP16—Or83b, LexA::GAD—GH146, LexA—Repo, LexA—OK107, All strains: (73)	LexAop—rCD2::GFP (73), LexAop—G-CaMP1.6 (102)
QF—QUAS	*Neurospora crassa*	(77)	QS (repression can be relieved by quinic acid)	QF-GH146, Others (e.g., http://flystocks.bio.indiana.edu)	QUAS—mCD8::GFP, QUAS—mtdTomato::3xHA, QUAS-nuclacZ, QUAS-shi^{ts1}, Others (e.g., http://flystocks.bio.indiana.edu)

of PNs by using GH146-GAL4 (69). Optical calcium imaging using G-CaMP 1.3 has also been performed in intrinsic mushroom body neurons by Wang et al. (70). Altogether, it has been demonstrated that GECIS can be used to visualize odor-evoked Ca^{2+} activity either in specific OSNs using specific OR-GAL4-lines (23, 33) or more broadly in the majority of OSNs using Orco-GAL4 (26), in PNs using GH146-GAL4 (69), and in intrinsic mushroom body neurons (Kenyon cells) using one out of many available GAL4-driver lines (51).

Since this initial success of establishing the GAL4-UAS system for calcium imaging in specific neuronal populations, a number of complementary tools have been developed based on the GAL4 repressor protein GAL80; the flip-out technique (71) and MARCM technique (mosaic analysis with a repressible cell marker (72)) allows a detailed lineage analysis of subsets of neuron populations or even single neurons targeted by the GAL4 driver line. GAL80^{ts}, a modified version of GAL80 allows for a temporal control of the system. At lower temperatures (~19°C) the GAL4 transcription activation is suppressed by binding of GAL80^{ts} to the GAD (GAL4-activation domain), whereas increasing the temperature releases GAL80^{ts} of the GAD and allows transcription activation of GAL4. The above listed methods all allow for restricting the expression of transgenes down to individual neurons.

Analogous to the GAL4 system, Lai and Lee (73) developed an alternative binary transcriptional system, called *LexA-LexAop system*. This tool is based on a bacterial expression system (*Escherichia coli*) and can be employed in addition and simultaneously to the GAL4-UAS system, and thus allows a more versatile analysis of diverse cell types. The effector gene LexA (74) comes in two different versions: it is either fused to the viral VP16 domain (75) to initiate transcription or the GAD domain of GAL4, which can be inhibited by GAL80. This implies that the system can either be employed to yield a complete background of an additional neuronal population labeled (LexA::VP16, GAL80 unaffected) or be used for MARCM analysis in parallel to the GAL4-UAS MARCM technique (dual-expression-control MARCM (76)).

Recently the development of a third binary transcriptional system has been reported. The *Q-system* employs an expression system originating from the red bread mold *Neurospora crassa* (77). It involves the transcription activator QA-1F (QF), which interacts with five repeats of a 16 bp sequence (QUAS) to initiate the expression of any reporter gene. Although no QUAS-GECI line is available at the time of writing, it will very likely be engineered in the near future. The advantage of the Q system is that it comes with its own GAL4 independent repressor, namely QA-1S (QS), enabling a repression of QF activity independent of GAL80 and thus, unlike the LexA system, a GAL4 independent MARCM technique ("independent double MARCM", for details see ref. (77)) is possible.

Another neat extension is that QS repression can be relieved by the addition of quinic acid to the food, which allows for a temporal control of QF activity.

Although the GAL4-UAS-system still represents the most commonly used binary expression system for *Drosophila*, it is likely that these recent developments can be used to express GECIs and, therefore, allow for an even more restricted expression of the GECI to particular neurons of interest. In the following we will describe which technical equipment is necessary for performing optical calcium imaging in the *Drosophila* olfactory system.

2.2. Components of the Imaging Setup Using a Widefield Microscope

– A fixed stage upright widefield fluorescence microscope (e.g., Axioskop 2FS or Axio Examiner from Zeiss or BX51WI from Olympus) is favorable. It should be equipped with a 20 and/or 40× water immersion objective (e.g., Plan-Apochromat, Zeiss or XLUM Plan FI, Olympus).

– As a light source either a filter-equipped Xenon lamp (Lambda DG-4, Sutter Instruments) or a monochromator (e.g., VisiChrome, Visitron Systems or Polychrome V, TILL Photonics) can be used. For exciting FRET-based sensor proteins we use an excitation wavelength of ~440 nm, for exciting G-CaMP a wavelength of ~480 nm can be used.

– For separating excitation wavelength from emission wavelength a long pass filter (LP) has to be used. Thus for FRET-based GECIs a 455 nm LP filter and for single-chromophore GECIs a 500 nm LP filter is recommended.

– Cooled CCD camera (e.g., Coolsnap HQ2 or Cascade II, Photometrics or PCO Imaging, Sensicam). The two emission wavelengths of FRET-based sensors have to be recorded separately. A beamsplitter device that projects the images of the two emission wavelengths onto the two halves of the CDD camera chip using appropriate dichroic mirrors and emission filters for YFP and CFP is available from Photometrics (Dual View, Photometrics, Tucson, AZ).

– Software for data acquisition and data analysis (e.g., the software MetaFluor for data acquisition and MetaMorph (both Visitron Systems) or custom written IDL software (ITT Visual Information Solutions) for data analysis).

2.3. Components of the Imaging Setup Using a Two-Photon Laser Scanning Microscope

Two-photon excitation processes were investigated for several decades. The important work by Denk et al. (78) launched a new revolution in nonlinear optical microscopy. These days complete two-photon microscopy systems are available from several companies, e.g., Zeiss (LSM 7 MP), Leica (TCS MP5), or Prairie Technologies. Here we list briefly the most important components that are needed to set up a two-photon microscope.

- An upright microscope (e.g., Axio Imager or Axio Examiner from Zeiss) with a high-precision Z-drive and a motorized XY-scanning stage. It should be equipped with a 20, 40 and/ or 60× water immersion objective with high numerical aperture (e.g., Plan-Apochromat-series from Zeiss).

- X/Y-scanning unit with high scanning speed (ca. 5 frames/sec with 512×512 pixel resolution) and bidirectional scan property. A freely rotating scan head (360°) is advisable.

- An ultrafast mode-locked titanium:sapphire laser for the wavelength range 700–1,000 nm as a light source (e.g., Chameleon lasers from Coherent or Mai Tai/Tsunami lasers from Spectra-Physics) with dichroic mirror (595DCXR for G-CaMP) allowing infrared laser beam to pass and reflect visible photons and rejection filter (690LP for G-CaMP). The laser beam has to be coupled into the microscope either directly using relay mirrors or through fiber coupling. The beam path needs to be enclosed to prevent accidentally exposing the operator to the beam. Excitation laser light of 925 nm wavelength is recommended for imaging using G-CaMP.

- High-sensitivity photomultiplier (PMT) as a detector.

- Microscope and laser should be mounted on a vibration-isolation table (well-damped, 8-inch-thick tables mounted on air supports, e.g., from Newport).

- Software for data acquisition (e.g., ZEN for the LSM710 NLO microscope, Zeiss). Data analysis needs to be done with a separate program (e.g., custom written IDL software (ITT Visual Information Solutions) or ImageJ including corresponding plug-ins).

2.4. Materials for the In Vivo Preparation

We are describing in the next paragraph two alternative methods (Method A and B) that can be used for the in vivo fly brain dissection for optical recording, both of which have particular advantages (Fig. 2). Depending on the dissection method chosen, different materials are needed as listed below:

Method A

- 1 ml pipette tips.
- Plastic cover slips (Plano, Wetzlar, Germany).
- Very thin transparency.
- Blade breaker and breakable razor blades (Fine Science Tools).
- Very fine forceps (e.g., Dumont # 5, Fine Science Tools).
- Harmless and odor-free glue (e.g., the dental glue Protemp II, ESPE 3M).
- Modeling clay.

54   A. Strutz et al.

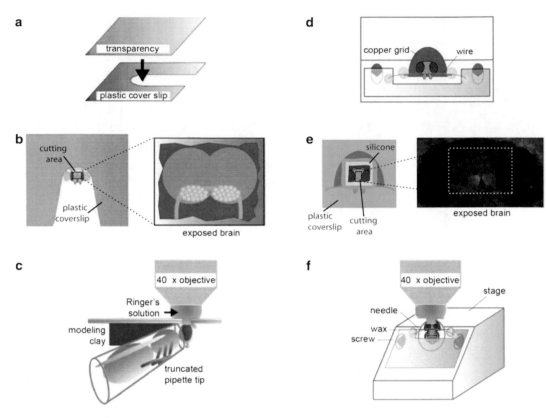

Fig. 2. In vivo preparation for optical calcium imaging of olfactory neurons in *Drosophila* using method A (**a–c**) or method B (**d–f**). (**a**) A notch is cut into a plastic cover slip onto which a very thin transparency is glued. (**b**) The fly's head can then be located under the transparency so that a window can be cut through the transparency and into the head capsule, thereby exposing the brain. (**c**) To position the head in the right angle under the transparency the fly can be placed into a truncated 1 ml-pipette tip so that the head is sticking out. The restrained fly can be positioned using modeling clay and the head glued under the transparency using odor-free dental glue. The preparation is then positioned under the microscope equipped with a water immersion objective. (**d**) Flies are placed into a custom-made Plexiglas mounting block by fixing the head with a copper plate. The antennae are pulled forward with a fine metal wire to allow optical access to the ALs. (**e**) A polyethylene foil is attached on the head and sealed to the cuticle with two-component silicone. A small window is cut through the foil and cuticle to expose the brain. (**f**) The mounting block is placed under the microscope so that the ALs are accessible for imaging using a water immersion objective; **d–f** from Veit Grabe (Max Planck Institute, Jena).

- Ringer's solution containing 130 mM NaCl, 5 mM KCl, 2 mM MgCl ($\times 6H_2O$), 2 mM CaCl$_2$ ($\times 2H_2O$), 36 mM saccharose, 5 mM Hepes, pH 7.3 (79).

Method B

- Custom-made Plexiglas mounting block with copper plate (Athene Grids, Plano) and built-in screws.
- Insect pins (Fine Science Tools).
- Rosin dissolved in ethanol (Royal Oak Rosinio, Royal Oak, Germany).
- Plastic cover slips (Plano).
- Thin polyethylene foil.

- Fine metal wire (HP Reid co. inc., www.hpreid.com).
- Sapphire blade (WPI, www.wpiinc.com).
- Very fine forceps (e.g., Dumont # 5, Fine Science Tools).
- Beeswax.
- Two-component silicone (KwikSil, WPI, www.wpiinc.com).
- Ringer's solution (see above (79)).

3. Procedures

3.1. Dissection for In Vivo Imaging

3.1.1. Method A (Modified from Ref. (80))

Flies are briefly immobilized on ice and carefully sucked into a truncated blue pipette tip so that the head and half of the thorax are sticking out. A notch is cut into a plastic cover slip and the thin transparency glued onto it (Fig. 2a). The pipette tip with the fly inside is positioned under the transparency and fixed using modeling clay. The head of the fly is then glued under the thin transparency using the dentist glue. Attention should be paid to keep the antennae free of glue and dry. A hole is then cut through the transparency and through the head capsule using a splint of a razor blade and a blade holder to expose the brain (Fig. 2b). During the preparation the brain should be covered with Ringer's solution (see above). Using forceps tracheae are carefully removed without damaging the brain or the antennal nerves and the preparation is placed under the microscope (Fig. 2c). Overall, the entire preparation should last less than 10 min.

3.1.2. Method B (e.g., Ref. (81–83))

Flies are anesthetized on ice and fixed with the neck onto a Plexiglas mounting block using a copper plate and a small needle in front of the head, whereas thorax and abdomen are hanging (Fig. 2d). The head is glued to the stage with rosin (diluted in ethanol) and the antennae are pulled forward using a fine metal wire which is attached to a plastic coverslip and placed in the cuticular fold between the antenna and the head. The coverslip needs to be fixed to the front of the mounting block with beeswax. With little screws that are built into the mounting block, the coverslip with the metal wire should be pushed forward gently to allow subsequent access to the ALs. A second plastic coverslip with a hole that is covered with polyethylene foil is glued to the head with beeswax and sealed to the cuticle with two-component silicone (Fig. 2e). A small hole is cut through the foil and cuticle with a sapphire blade. Immediately after opening the head capsule, the brain needs to be covered with Ringer's solution (see above), which should be exchanged several times. With fine forceps tracheae and glands are carefully removed to allow optical access to the ALs and the mounting block containing the fly is placed under the microscope (Fig. 2f).

3.2. Odor Stimulation: Choosing the Appropriate Stimulus Device

Odor stimulus application is one of the most challenging tasks when performing optical recording experiments of olfactory neurons in the fly brain. The olfactory system of the fly is extremely sensitive to low odor concentrations due to the large number of OSNs as well as cognate expressed receptors and signal amplification (e.g., by electrically connected neurons). Moreover the diffusion of volatile substances is difficult to monitor. Depending on the question addressed, the odor concentration required defines the complexity of the stimulation device. For investigations using generally highly concentrated odors (e.g., 10^{-1} or 10^{-2} of pure odor diluted in a given solvent, e.g., water or mineral oil), contaminations are often comparatively low. However, low concentrations (e.g., 10^{-8}–10^{-3}) are often included in investigations of dose–response functions or for the identification of eligible ligands of ORs. Here, contamination is the most disturbing factor during the optical imaging recording. Since it is necessary to avoid residual contamination between two recordings, a highly accurate olfactometer, allowing spatially and temporally controlled odor delivery, must be employed.

Here we describe two types of olfactometers (Fig. 3). The stimulus device developed by Galizia and colleagues (34, 84) is fast and easily applicable. The second system, recently engineered by Olsson et al. (85) is a spatially and temporally accurate multicomponent system. Both systems allow for a computer-controlled odor application in a constant airstream as well as the application of two or more odors at the same time.

The "classical" stimulus device consists of a glass-pipe conducting a constant airstream to the animal's antennae (Fig. 3a). Into the main airstream an odorized airstream is added during stimulation: two Pasteur glass pipettes achieve additional access to the air-delivery pipe via small holes. One pipette is empty and thus continuously adds filtered air into the constant air stream in between stimulations. The second pipette contains a piece of filter paper (e.g., Whatman) soaked with 5–10 μl odor diluted in a solvent as for example mineral oil. Odor concentration is configured by specific dilutions of the odor in the solvent. During odor stimulation clean air delivery of the empty pipette is blocked. Airflow through the odor containing pipette is switched on simultaneously and thus sustains the main airflow to avoid mechanical artifacts. For each measurement the odor pipette can be replaced with a pipette containing a different odor. Both pipettes are connected via silicone tubing and magnetic valves to flowmeters that adjust the airflow. The flowmeters are in turn connected to a stimulus controller (Syntech), which regulates the airflow and interfaces with a computer to control the on- and offset of the measurements. The stimulus controller contains carbon filters to provide a clean airflow. For optimal odor delivery several parameters needs to be controlled accurately such as the diameter and length of the exit tube and the

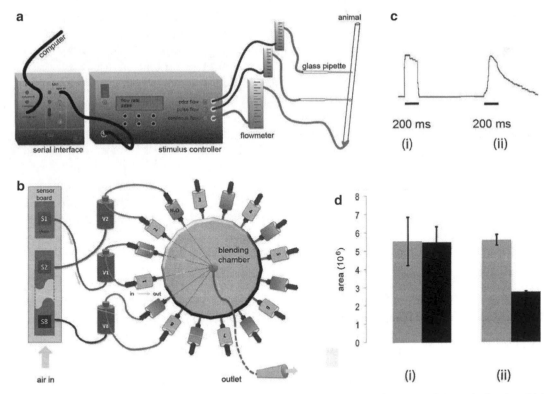

Fig. 3. Comparison of a classical olfactometer in comparison to a multicomponent olfactometer. Schematic drawing of (**a**) the classical olfactometer and (**b**) the multicomponent stimulus device. (**c**) Photoionization detector (PID) measurements using ethanol: (i) multicomponent stimulus device shows a square odor pulse, terminating immediately at the end of the pulse, (ii) the classical olfactometer shows a slight past-leakage, indicating odor delivery after pulse termination. (**d**) SPME (solid-phase microextraction) measurements of pure isoamyl acetate (*gray*) and pure 2-heptanone (*black*) show (I) an almost equal odor concentration for headspace stimulation with the multicomponent stimulus device with adjusted flow rates and (II) at equal flow rates. Adapted from (85).

position of the animal along the airstream (for detailed experimental descriptions see ref. (86)).

The olfactometer developed by Olsson et al. (85) (Fig. 3b) is a multicomponent stimulus device. In this case, airflow rates are used to equilibrate odor concentrations depending on the individual partial vapor pressures of the chemicals. The system consists of three major parts: the airflow regulator, an odorant delivery system and a blending chamber. Compressed air is separated into different channels and connected to a sensor board with a series of eight flowmeters (using choke valves). These control the separate airstreams downstream of the flowmeter according to the vapor pressure of the odorants. The individually controlled airstreams are fed into eight Teflon tubes and guided to the odorant delivery system. The delivery system consists of three-way solenoid valves that release the air through two distinct exit check valves: one line leads to a closed chamber made of PEEK (polyetheretherketone) containing water and producing humidified air. The second line leads to a chamber containing the odor, producing odorized air. Under non-stimulation conditions air is guided through the water-filled chamber and the sum of all eight humidified airstreams is combined to a constant air stream. During stimulation with a single odor, the cognate water check-valve is closed and the air is lead only through the equivalent odor chamber, thereby the maintaining a constant total airflow.

Both water and odor chambers are equipped with ball-stop check valves at air inlets and outlets to prevent leakage or contamination. This tight sealed system allows the odor chamber to fill with a saturated headspace during inter-stimulus intervals. All airflows are combined in a single concentric PEEK blending chamber and the unified airstream is guided via Teflon tubing to the animal's antennae. The distinct airflows necessary for adequate odor concentrations can be calculated using common gas laws (for detailed descriptions see ref. (85)).

Both systems allow odor concentration control via dilution in a solvent. In the classical system the odor vaporizes from filter paper. Chemical interactions with the adsorbent surface must therefore be considered as an influencing factor. Another important aspect is the so-called baseline degradation (86), which describes the gradual reduction of odor concentration over time entailed by repetitive airflow above the filter paper.

The multicomponent system collects saturated headspace from a closed odor vial. This strategy allows additional odor concentration control via airflow, but the interstimulus interval must last until the volatile odor can completely saturate the headspace. Comparative analysis using photoionization detector (PID) measurements made by Olsson et al. (85) shows a minor accuracy difference of the classical olfactometer in comparison to the multicomponent device (Fig. 3c). The PID measurement with ethanol reveals a small post-leakage after odor stimulation, which is caused by the open construction of the system compared to the tightly closed chambers of the multicomponent system.

Another comparison made by Olsson et al. (85) indicates the influence that equilibration of odor concentration via airflow can have. SPME (solid-phase micro extraction) measurements of pure isoamyl acetate (partial vapor pressure 5.68 mmHg at 25°C) and pure 2-heptanone (4.73 mmHg at 25°C) at calculated airflow settings using common gas laws (depending on partial vapor pressures) and with similar airflow rates (Fig. 3d) clearly show a difference between the number of odor molecules reaching the SPME fiber. The multicomponent device uses digital flowmeter control, which allows easy adjustment of individual airflows, whereas the classical olfactometer contains a single manual system to adjust a common airflow for all odors, although they could potentially be adjusted in between each stimulation.

A major advantage of the classical olfactometer is the easy exchangeability of odors simply by replacing the odor pipettes. It is therefore quite useful for studying response profiles of large odor sets. It is also sufficient for investigations of different odor concentrations as long as the concentrations vary at least about one order of magnitude in the solvent. The multicomponent device affiliates eight odor chambers solely, but it is in turn well suited for accurate investigations of blend interactions as well as precise dose response curves.

3.3. Imaging Data Acquisition

Using the widefield imaging setups described reasonable data acquisition can be performed at frame rates up to 10 Hz. However, the frame rate is dependent on signal strength as higher fame rates allow for less light detection per frame. In addition, light intensity should be adjusted such that bleaching does not affect the readout over the time course of the experiment. For single-chromophore GECIs one image per frame is acquired, whereas for FRET-based sensors two images for donor and acceptor are acquired. A ratio image of acceptor/donor emission intensity can be calculated after the acquisition process. The recording can last up to 2 h depending on the condition of the animal.

3.4. Imaging Data Analysis

It is advisable to use several corrections of the raw imaging data to allow for correct data interpretation. These corrections are listed and described briefly, but see Galizia and Vetter (87) for more detailed information regarding imaging data analysis.

Movement correction: In order to remove shifts in the XY-position of the recorded neurons/glomeruli during the measurement of one fly, the images within one measurement and between different measurements need to be aligned.

Data filtering: Functional imaging at a widefield microscope has the disadvantage in contrast to two-photon imaging that the fluorescent light from one point may scatter and be reflected back into the objective (see below for the comparison between single- and multiphoton imaging, Fig. 5). Hence a scattered light correction algorithm might be useful to reduce this type of distortion but has to be used with caution (see ref. (87) for step-by-step instructions). Another very important data correction method is the bleaching correction. Since the exposure to excitation light may lead to a fast decay of the fluorescent emission intensity by the sample during the measurement, it may be necessary to correct for bleaching before analyzing the data. The bleaching artifact is strongly correlated with the time of exposure and the inter-frame-interval. One method to correct for bleaching is to calculate for each single measurement a background frame (i.e., take the average of several frames just before stimulus application; F_0), to divide each frame by this background and to calculate the relative fluorescence changes ($\Delta F/F_0$). In order to subtract the fluorescence decay due to bleaching one can subsequently calculate the average relative fluorescence change of all pixels for each frame and fit a logarithmic function to this change. The fitted log-function from the relative fluorescence curve can then be subtracted for each pixel. These data can be transformed back to "raw" fluorescence data by multiplying the bleach corrected frames with the background.

Signal calculation: To achieve a comparable standard for the calculation of the relative fluorescence changes ($\Delta F/F$) for single-chromophore sensors (e.g., G-CaMP), the fluorescence background has to be first subtracted from the averaged values of frames just before the stimulus application (i.e., calculation of background frame: F_0) in each measurement, so that the basal fluorescence is normalized to zero. Then calculate the relative fluorescence changes by dividing the background and multiplying it with 100 to obtain percentage fluorescence changes (i.e., $\Delta F/F = (F_i - F_0)/F_0 \times 100$; where F_0 is the background frame and F_i the fluorescence value for the ith frame of the measurement). Regarding FRET-based sensors (e.g., Cameleon) the ratio of double emission measurements has to be calculated (e.g., for Cameleon: $F_{Ratio} = F_{YFP}/F_{YCFP}$). If both wavelengths show the same bleaching decay, the bleaching artifact is already corrected by calculating the ratio. However, if both wavelengths show unequal bleaching, a bleaching correction needs to be performed independently for each wavelength following the procedure as described above.

Signal representation: The fluorescence change can be best represented as false-color coded images by subtracting a frame during the stimulus application from a frame just before application. The temporal properties of the fluorescence changes can be shown as time traces from single measurements. Plot the relative fluorescence changes ($\Delta F/F_0$) from different glomeruli/neurons over time by calculating the average $\Delta F/F_0$ within a region of interest (square or circle) with a diameter of 10 μm corresponding to the size of a glomerulus in the fly AL. It is also favorable to calculate the peak amplitude of the relative fluorescence change to obtain just one value as the Ca^{2+} signal for each glomerulus/neuron during the odor application. The peak response can also be quantified by integrating the area below the time trace during the stimulus application. If the glomerular structures are clearly visible, the Ca^{2+} responses can afterwards be mapped to identified glomeruli using the 3D AL atlas by Laissue et al. (88). This procedure allows for comparisons between different individuals.

3.5. A Comparison of FRET-Based and Single-Chromophore Sensors

Either FRET-based or single-chromophore GECIs may be more or less favorable for monitoring odor-evoked Ca^{2+} transients in olfactory neurons of the *Drosophila* brain, depending on the particular application and imaging method. We have expressed a variety of GECIs in a large population of olfactory PNs using the driver line GH146-GAL4 (69). Using a widefield microscope we have focused on terminal arborizations of PNs in the calyx region of the mushroom body (Fig. 4). Odor-evoked Ca^{2+} activity has been monitored using the FRET-based sensor proteins Cameleon 2.1 (68), Cameleon 6.1 (89), TN-XL (90) and the single-chromophore sensor G-CaMP 1.6 (91). It can be seen that all GECIs are in principle appropriate to detect odor-evoked Ca^{2+} transients, with slight differences in the time course of the signal onset and the decay of the Ca^{2+} transient, which might be due to different expression levels and different dissociation constants of the particular GECI for Ca^{2+}. Note that a high noise caused by slight movements of the preparation is diminished in ratiometric, FRET-based sensors. Also note that the G-CaMP 1.6 sensor shows a comparatively strong bleaching at the given optical conditions. However, the single-chromophore sensor G-CaMP shows comparatively strong signal intensity. It must be noted, though, that more recently published sensor proteins might offer advantages in terms of signal intensity and signal-to-noise ratio. In particular, the FRET-based sensors Cameleon 3.6 (92) and TN-XXL (93), and the single-chromophore sensor G-CaMP 3.0 (94) have been described to offer strong improvements in this regard (62, 93, 94).

3.6. A Comparison of Widefield and Two-Photon Functional Imaging

Two-photon microscopy is a powerful tool that combines scanning microscopy with two-photon excitation to create high-resolution, three dimensional images of microscopic samples. Compared to widefield fluorescent microscopy techniques, two-photon imaging causes less photo damage to living cells and the fluorophores, permits deeper tissue penetration and offers inherent 3D optical sectioning (95). Moreover, the two-photon excitation is restricted to the plane of focus allowing for imaging of small structures with high signal-to-noise ratio and, therefore, improved background discrimination. Hence two-photon microscopy is used in many fields of biomedical research, where high spatial and temporal resolution of deep living tissue is of particular interest. However, widefield functional imaging is still a useful technique and offers several advantages. One important advantage is that the imaging setup is relatively simple and affordable. It can be used without any special skills and needs only little maintenance. A two-photon system requires that the operator has a good background in physics since it needs a lot of constant care and costly maintenance. In order to compare both systems, we have performed optical calcium imaging experiments and monitored the odor-evoked Ca^{2+} responses of a fly AL expressing the calcium sensor G-CaMP in OSNs using widefield and two-photon functional imaging (Fig. 5). The comparison shows that the strongly activated glomeruli to a specific odor can be individually identified using both systems. Widefield microscopy enables one to yield a quick overview about neuronal activity to a specific odor in a considerable large number of glomeruli just with a single measurement. However, the two-photon

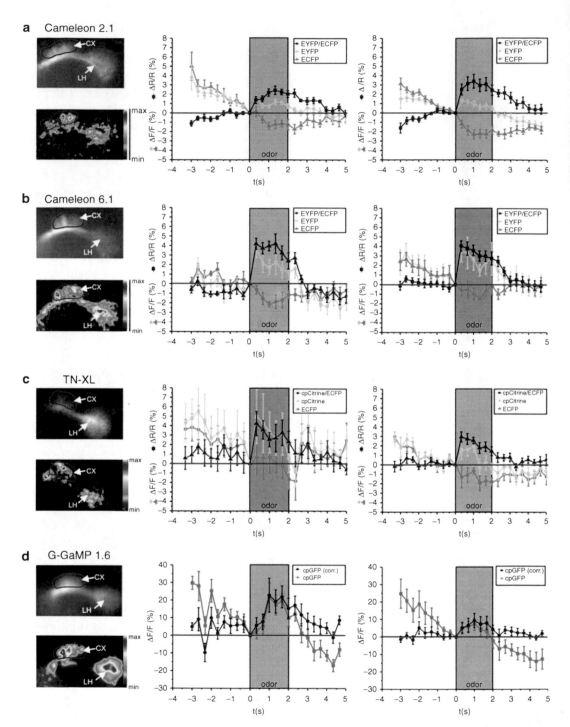

Fig. 4. Odor-evoked Ca²⁺ dynamics in terminal arborizations of the PNs using different calcium sensors. On the *left row* the baseline fluorescence of the sensor protein expressed indicates the anatomy of the terminal arborizations. In the case of FRET-based sensors (**a–b**) the acceptor fluorescence is depicted. Below the anatomical image a *false-color coded* image indicates regions of high and low Ca²⁺ increase in response to an odor stimulus. *Warm colors* represent regions of high Ca²⁺ activity, *cold colors* regions of low or no Ca²⁺ increase. The *red* region encircling the calyx indicates the area in which the temporal course of Ca²⁺ increase shown in the *middle row* is measured. The *middle row* indicates five repetitive stimulations with the odor in the same individual animal. In the *right row* the average of five animals is shown. Data represent means ± SEM.

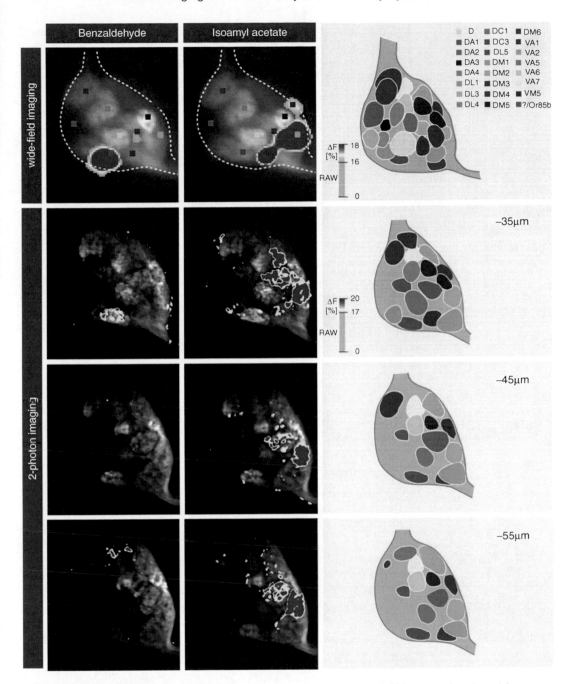

Fig. 5. Comparison of widefield and multiphoton functional imaging. Activity of OSNs expressing the calcium sensor G-CaMP are imaged during the application of two different odors (benzaldehyde and isoamyl acetate) using either a widefield (above) or a multiphoton imaging system (below). Values below the $\Delta F/F$ threshold (scale to the right) are omitted to illustrate the specificity of the signals, as well as the glomerular arrangement as visualized by the intrinsic fluorescence. Multiphoton imaging allowed to image three different focal planes of the AL and thus visualizing the activity of different sets of glomeruli. Different schematized atlases of the AL representing the glomeruli that have been functionally imaged are shown to the right. Glomeruli identity is color-coded.

system offers more information due to the capability of deep optical sectioning through different layers of glomeruli. Moreover, the Ca^{2+} signals are clearly restricted to single glomeruli, whereas the widefield imaging setup provides only images with broad odor responses due to light scattering. Nevertheless it should be kept in mind that the probability to overlook neuronal activity is higher at the two-photon system because the z-level size imaged is solely 1 μm. Hence it is recommended to perform a complex 3D scan, which is a challenging task due to limitations in movement and time.

3.7. Optical Calcium Imaging at Different Levels of Olfactory Processing

Using functional imaging the detection of the spatiotemporal distribution of odor-evoked Ca^{2+} transients is possible at various levels of olfactory processing. Dependent on the GAL4 driver line a GECI can be expressed in OSNs, PNs or KCs, respectively. The example depicted in Fig. 6 shows expression of the sensor Cameleon 2.1 (68). The left column shows the baseline fluorescence of the EYFP chromophore. The middle column represents the spatial distribution of Ca^{2+} activity evoked by the odor stimulus and the right column the temporal dynamics of Ca^{2+} transients detected by Cameleon 2.1. In conclusion, at all five levels of processing odor-evoked Ca^{2+} transients can be reliably detected.

4. Notes

Optical calcium imaging in the brain of a living animal is not a simple task, and problems might arise at various steps of the procedure. For setting up the technique properly we recommend to particularly take care of the following points.

4.1. Intact Preparation

Of course, successful and meaningful experiments require that the animals prepared such that the brain and the olfactory organs are kept intact. In particular, attention should be paid that the antennal nerve remains untouched and the antennae and maxillary palps dry.

4.2. Movement Artifacts

It is very important to fix the fly and stabilize the brain in the best possible way during the dissection procedure to avoid strong movements since these will cover your fluorescence changes. Typical

Fig. 6. (continued) *Left row* (**a–e**): EYFP baseline fluorescence in olfactory neurons. Scale bars: 25 μm. *Middle row* (**a′–e′**): false-color coded illustration of the spatial distribution of odor-evoked Ca^{2+} activity in the neurons labeled in the *left row*, as indicated by the *white outlines*. *Warm colors* represent regions of high Ca^{2+} activity, *cold colors* regions of low or no Ca^{2+} increase. *Right row* (**a″–e″**): temporal dynamics of Ca^{2+} activity in regions of high Ca^{2+} activity, indicated as *black circles* in the *middle row*. For OSNs (**a″**) the relative changes in EYFP and ECFP emission are indicated in addition to the ratio EYFP/ECFP to demonstrate the change in emission intensities in opposite directions. *Traces* indicate means ± SEM of 3–6 odor stimulations within the same animal. The *gray bars* indicate the duration of the 2 s odor stimulus.

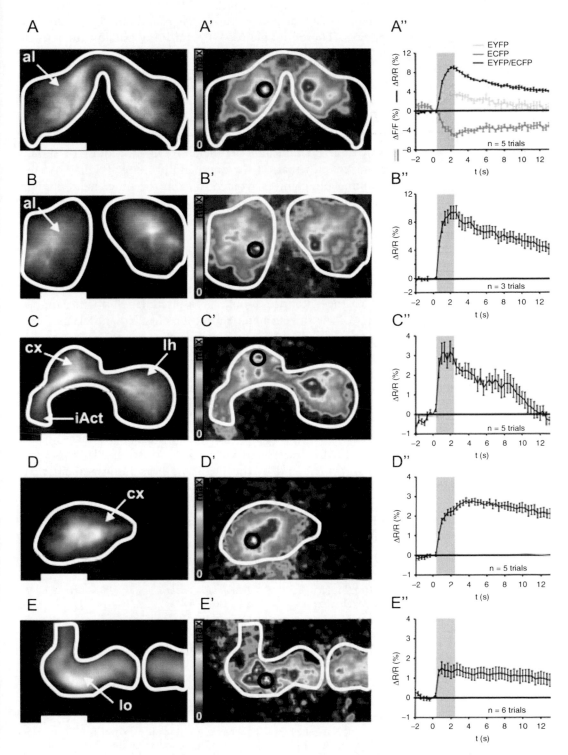

Fig. 6. Optical imaging of odor-evoked Ca²⁺ dynamics in first, second and third order olfactory neurons using Cameleon 2.1. Odor-evoked Ca²⁺ activity is monitored in OSNs within the AL (al) **(a–a″)**, in olfactory PNs arborizing with the AL **(b–b″)**, in terminal arborizations of PNs in the mushroom body calyx (cx) and the lateral horn (lh) **(c–c″)**, in intrinsic mushroom body neurons at the level of the calyx **(d–d″)** and in the main output region of the mushroom bodies, the lobes (lo) **(e–e″)**.

movement artifacts arise from moving legs and muscles around the oesophagus. Therefore, we recommend immobilizing the legs by gluing them with bees wax to the mounting block as well by fixing the proboscis with a fine needle (dissection Method B).

In addition it is advisable to remove the oesophagus or to cut the muscle bundles next to the oesophagus (dilators of the pharynx) to constrain it from movements.

4.3. Optimal Illumination Conditions

Exposure time and frame rate (frequency) of image acquisition should be optimized such that on the one hand prolonged light exposure and resulting bleaching is avoided, but on the other hand enough light is capture for each frame and signal-to-noise is sufficient.

4.4. Odor Application

The most challenging part in doing experiments with odors is to avoid odor contamination. The above mentioned olfactometers provide solutions to avoid that, but of course, care should be taken. The room should be as odor-free as possible, and odor streams applied to the antennae should be aspirated behind the animal. The parts of the olfactometer should be checked and cleaned frequently to avoid potential contamination.

Acknowledgments

This work was supported by the Federal Ministry of Education and Research (BMBF research group to S.S.), the Max Planck Society (to S.S. and A.S.), and by the Deutsche Forschungsgemeinschaft (GRK 1156 to T.V., SFB 554 to A.F. and T.R., SPP 1392 "Integrative Analysis of Olfaction" to S.S. and A.F.). We are grateful to Erich Buchner and Bill S. Hansson for support.

References

1. Vosshall LB, Stocker RF (2007) Molecular architecture of smell and taste in *Drosophila*. Annu Rev Neurosci 30:505–533

2. Liang L, Luo L (2010) The olfactory circuit of the fruit fly *Drosophila melanogaster*. Sci China Life Sci 53:472–484

3. Arora K, Rodrigues V, Joshi S, Shanbhag S, Siddiqi O (1987) A gene affecting the specificity of the chemosensory neurons of *Drosophila*. Nature 330:62–63

4. Ayer RK Jr, Carlson J (1991) acj6: a gene affecting olfactory physiology and behavior in *Drosophila*. Proc Natl Acad Sci USA 88: 5467–5471

5. Ayyub C, Paranjape J, Rodrigues V, Siddiqi O (1990) Genetics of olfactory behavior in *Drosophila melanogaster*. J Neurogenet 6: 243–262

6. Carlson J (1991) Olfaction in *Drosophila*: genetic and molecular analysis. Trends Neurosci 14:520–524

7. Stocker RF, Lienhard MC, Borst A, Fischbach KF (1990) Neuronal architecture of the antennal lobe in *Drosophila melanogaster*. Cell Tissue Res 262:9–34

8. Stocker RF, Singh RN, Schorderet M, Siddiqi O (1983) Projection patterns of different types of antennal sensilla in the antennal glomeruli of *Drosophila melanogaster*. Cell Tissue Res 232:237–248

9. de Bruyne M, Clyne PJ, Carlson JR (1999) Odor coding in a model olfactory organ: the

Drosophila maxillary palp. J Neurosci 19: 4520–4532

10. de Bruyne M, Foster K, Carlson JR (2001) Odor coding in the *Drosophila* antenna. Neuron 30:537–552

11. Hallem EA, Carlson JR (2006) Coding of odors by a receptor repertoire. Cell 125: 143–160

12. Hallem EA, Ho MG, Carlson JR (2004) The molecular basis of odor coding in the *Drosophila* antenna. Cell 117:965–979

13. Bhandawat V, Olsen SR, Gouwens NW, Schlief ML, Wilson RI (2007) Sensory processing in the *Drosophila* antennal lobe increases reliability and separability of ensemble odor representations. Nat Neurosci 10:1474–1482

14. Chou Y-H, Spletter ML, Yaksi E, Leong JCS, Wilson RI, Luo L (2010) Diversity and wiring variability of olfactory local interneurons in the Drosophila antennal lobe. Nat Neurosci 13:439–449

15. Olsen SR, Bhandawat V, Wilson RI (2007) Excitatory interactions between olfactory processing channels in the *Drosophila* antennal lobe. Neuron 54:89–103

16. Wilson RI, Turner GC, Laurent G (2004) Transformation of olfactory representations in the *Drosophila* antennal lobe. Science 303: 366–370

17. Seki Y, Rybak J, Wicher D, Sachse S, Hansson BS (2010) Physiological and morphological characterization of local interneurons in the *Drosophila* antennal lobe. J Neurophysiol 104:1007–1019

18. Miyawaki A (2003) Fluorescence imaging of physiological activity in complex systems using GFP-based probes. Curr Opin Neurobiol 13:591–596

19. Shanbhag SR, Mueller B, Steinbrecht RA (2000) Atlas of olfactory organs of *Drosophila melanogaster*. 2. Internal organization and cellular architecture of olfactory sensilla. Arthr Struct Dev 29:211–229

20. Shanbhag SR, Singh K, Singh RN (1995) Fine structure and primary sensory projections of sensilla located in the sacculus of the antenna of *Drosophila melanogaster*. Cell Tissue Res 282:237–249

21. Clyne PJ, Warr CG, Freeman MR, Lessing D, Kim J, Carlson JR (1999) A novel family of divergent seven-transmembrane proteins: candidate odorant receptors in *Drosophila*. Neuron 22:327–338

22. Couto A, Alenius M, Dickson BJ (2005) Molecular, anatomical, and functional organization of the *Drosophila* olfactory system. Curr Biol 15:1535–1547

23. Fishilevich E, Vosshall LB (2005) Genetic and functional subdivision of the *Drosophila* antennal lobe. Curr Biol 15:1548–1553

24. Vosshall LB (2001) The molecular logic of olfaction in *Drosophila*. Chem Senses 26: 207–213

25. Vosshall LB, Amrein H, Morozov PS, Rzhetsky A, Axel R (1999) A spatial map of olfactory receptor expression in the *Drosophila* antenna. Cell 96:725–736

26. Larsson MC, Domingos AI, Jones WD, Chiappe ME, Amrein H, Vosshall LB (2004) Or83b encodes a broadly expressed odorant receptor essential for *Drosophila* olfaction. Neuron 43:703–714

27. Benton R, Sachse S, Michnick SW, Vosshall LB (2006) Atypical membrane topology and heteromeric function of *Drosophila* odorant receptors in vivo. PLoS Biol 4:e20

28. Sato K, Pellegrino M, Nakagawa T, Nakagawa T, Vosshall LB, Touhara K (2008) Insect olfactory receptors are heteromeric ligand-gated ion channels. Nature 452:1002–1006

29. Abuin L, Bargeton B, Ulbrich MH, Isacoff EY, Kellenberger S, Benton R (2011) Functional architecture of olfactory ionotropic glutamate receptors. Neuron 69:44–60

30. Benton R, Vannice KS, Gomez-Diaz C, Vosshall LB (2009) Variant ionotropic glutamate receptors as chemosensory receptors in *Drosophila*. Cell 136:149–162

31. Wicher D, Schafer R, Bauernfeind R, Stensmyr MC, Heller R, Heinemann SH, Hansson BS (2008) *Drosophila* odorant receptors are both ligand-gated and cyclic-nucleotide-activated cation channels. Nature 452: 1007–1011

32. Gao Q, Yuan B, Chess A (2000) Convergent projections of *Drosophila* olfactory neurons to specific glomeruli in the antennal lobe. Nat Neurosci 3:780–785

33. Vosshall LB, Wong AM, Axel R (2000) An olfactory sensory map in the fly brain. Cell 102:147–159

34. Fiala A, Spall T, Diegelmann S, Eisermann B, Sachse S, Devaud JM, Buchner E, Galizia CG (2002) Genetically expressed Cameleon in *Drosophila melanogaster* is used to visualize olfactory information in projection neurons. Curr Biol 12:1877–1884

35. Wang JW, Wong AM, Flores J, Vosshall LB, Axel R (2003) Two-photon calcium imaging reveals an odor-evoked map of activity in the fly brain. Cell 112:271–282

36. Marin EC, Jefferis GSXE, Komiyama T, Zhu H, Luo L (2002) Representation of the glomerular olfactory map in the *Drosophila* brain. Cell 109:243–255

37. Wong AM, Wang JW, Axel R (2002) Spatial representation of the glomerular map in the *Drosophila* protocerebrum. Cell 109:229–241

38. Shang Y, Claridge-Chang A, Sjulson L, Pypaert M, Miesenböck G (2007) Excitatory local circuits and their implications for olfactory processing in the fly antennal lobe. Cell 128:601–612

39. Yaksi E, Wilson RI (2010) Electrical coupling between olfactory glomeruli. Neuron 67: 1034–1047

40. Huang J, Zhang W, Qiao W, Hu A, Wang Z (2010) Functional connectivity and selective odor responses of excitatory local interneurons in *Drosophila* antennal lobe. Neuron 67: 1021–1033

41. Dacks AM, Green DS, Root CM, Nighorn AJ, Wang JW (2009) Serotonin modulates olfactory processing in the antennal lobe of *Drosophila*. J Neurogenet 23:366–377

42. Busch S, Selcho M, Ito K, Tanimoto H (2009) A map of octopaminergic neurons in the *Drosophila* brain. J Comp Neurol 513: 643–667

43. Carlsson MA, Diesner M, Schachtner J, Nässel DR (2010) Multiple neuropeptides in the *Drosophila* antennal lobe suggest complex modulatory circuits. J Comp Neurol 518: 3359–3380

44. Yu D, Ponomarev A, Davis RL (2004) Altered representation of the spatial code for odors after olfactory classical conditioning; memory trace formation by synaptic recruitment. Neuron 42:437–449

45. Lin HH, Lin CY, Chiang AS (2007) Internal representations of smell in the *Drosophila* brain. J Biomed Sci 14:453–459

46. Tanaka NK, Awasaki T, Shimada T, Ito K (2004) Integration of chemosensory pathways in the *Drosophila* second-order olfactory centers. Curr Biol 14:449–457

47. Busto GU, Cervantes-Sandoval I, Davis RL (2010) Olfactory learning in *Drosophila*. Physiology 25:338–346

48. Fiala A (2007) Olfaction and olfactory learning in *Drosophila*: recent progress. Curr Opin Neurobiol 17:720–726

49. Heisenberg M (2003) Mushroom body memoir: from maps to models. Nat Rev Neurosci 4:266–275

50. Keene AC, Waddell S (2007) *Drosophila* olfactory memory: single genes to complex neural circuits. Nat Rev Neurosci 8:341–354

51. Aso Y, Grübel K, Busch S, Friedrich AB, Siwanowicz I, Tanimoto H (2009) The mushroom body of adult *Drosophila* characterized by GAL4 drivers. J Neurogenet 23:156–172

52. Luo SX, Axel R, Abbott LF (2010) Generating sparse and selective third-order responses in the olfactory system of the fly. Proc Natl Acad Sci 107:10713–10718

53. Turner GC, Bazhenov M, Laurent G (2008) Olfactory representations by *Drosophila* mushroom body neurons. J Neurophysiol 99: 734–746

54. Rodrigues V (1988) Spatial coding of olfactory information in the antennal lobe of *Drosophila melanogaster*. Brain Res 453:299–307

55. Rodrigues V, Buchner E (1984) (3H)2-deoxyglucose mapping of odor-induced neuronal activity in the antennal lobes of *Drosophila melanogaster*. Brain Res 324:374–378

56. Martin J-R, Rogers KL, Chagneau C, Brûlet P (2007) In vivo bioluminescence imaging of Ca^{2+} signalling in the brain of *Drosophila*. PLoS One 2:e275

57. Miyawaki A, Llopis J, Heim R, McCaffery JM, Adams JA, Ikura M, Tsien RY (1997) Fluorescent indicators for Ca^{2+} based on green fluorescent proteins and calmodulin. Nature 388:882–887

58. Romoser VA, Hinkle PM, Persechini A (1997) Detection in living cells of Ca^{2+}-dependent changes in the fluorescence emission of an indicator composed of two green fluorescent protein variants linked by a calmodulin-binding sequence. J Biol Chem 272:13270–13274

59. Nakai J, Ohkura M, Imoto K (2001) A high signal-to-noise $Ca^{(2+)}$ probe composed of a single green fluorescent protein. Nat Biotechnol 19:137–141

60. Hasan MT, Friedrich RW, Euler T, Larkum ME, Giese G, Both M, Duebel J, Waters J, Bujard H, Griesbeck O, Tsien RY, Nagai T, Miyawaki A, Denk W (2004) Functional fluorescent Ca^{2+} indicator proteins in transgenic mice under TET control. PLoS Biol 2:e163

61. Pologruto TA, Yasuda R, Svoboda K (2004) Monitoring neural activity and (Ca^{2+}) with genetically encoded Ca^{2+} indicators. J Neurosci 24:9572–9579

62. Hendel T, Mank M, Schnell B, Griesbeck O, Borst A, Reiff DF (2008) Fluorescence changes of genetic calcium indicators and OGB-1 correlated with neural activity and calcium in vivo and in vitro. J Neurosci 28:7399–7411

63. Reiff DF, Ihring A, Guerrero G, Isacoff EY, Joesch M, Nakai J, Borst A (2005) In vivo performance of genetically encoded indicators of neural activity in flies. J Neurosci 25: 4766–4778

64. Kamikouchi A, Wiek R, Effertz T, Gopfert MC, Fiala A (2010) Transcuticular optical imaging of stimulus-evoked neural activities in the

Drosophila peripheral nervous system. Nat Protoc 5:1229–1235

65. Kamikouchi A, Inagaki HK, Effertz T, Hendrich O, Fiala A, Gopfert MC, Ito K (2009) The neural basis of *Drosophila* gravity-sensing and hearing. Nature 458:165–171

66. Fischer JA, Giniger E, Maniatis T, Ptashne M (1988) GAL4 activates transcription in *Drosophila*. Nature 332:853–856

67. Brand AH, Perrimon N (1993) Targeted gene expression as a means of altering cell fates and generating dominant phenotypes. Development 118:401–415

68. Miyawaki A, Griesbeck O, Heim R, Tsien RY (1999) Dynamic and quantitative Ca^{2+} measurements using improved Cameleons. Proc Natl Acad Sci 96:2135–2140

69. Stocker RF, Heimbeck G, Gendre N, de Belle JS (1997) Neuroblast ablation in *Drosophila* P(GAL4) lines reveals origins of olfactory interneurons. J Neurobiol 32:443–456

70. Wang Y, Guo HF, Pologruto TA, Hannan F, Hakker I, Svoboda K, Zhong Y (2004) Stereotyped odor-evoked activity in the mushroom body of *Drosophila* revealed by green fluorescent protein-based Ca^{2+} imaging. J Neurosci 24:6507–6514

71. Golic KG, Lindquist S (1989) The FLP recombinase of yeast catalyzes site-specific recombination in the *Drosophila* genome. Cell 59:499–509

72. Lee T, Luo L (1999) Mosaic analysis with a repressible cell marker for studies of gene function in neuronal morphogenesis. Neuron 22:451–461

73. Lai S-L, Lee T (2006) Genetic mosaic with dual binary transcriptional systems in *Drosophila*. Nat Neurosci 9:703–709

74. Brent R, Ptashne M (1985) A eukaryotic transcriptional activator bearing the DNA specificity of a prokaryotic repressor. Cell 43:729–736

75. Triezenberg SJ, Kingsbury RC, McKnight SL (1988) Functional dissection of VP16, the trans-activator of herpes simplex virus immediate early gene expression. Genes Dev 2:718–729

76. Lai S-L (2007) Neural diversity in the *Drosophila* olfactory circuitry: a dissertation. GSBS dissertations, University of Massachusetts Medical School

77. Potter CJ, Tasic B, Russler EV, Liang L, Luo L (2010) The Q system: a repressible binary system for transgene expression, lineage tracing, and mosaic analysis. Cell 141:536–548

78. Denk W, Strickler J, Webb W (1990) Two-photon laser scanning fluorescence microscopy. Science 248:73–76

79. Estes PS, Roos J, van der Bliek A, Kelly RB, Krishnan KS, Ramaswami M (1996) Traffic of dynamin within individual *Drosophila* synaptic boutons relative to compartment-specific markers. J Neurosci 16:5443–5456

80. Fiala A, Spall T (2003) In vivo calcium imaging of brain activity in *Drosophila* by transgenic Cameleon expression. Sci STKE 174:l6

81. Pelz D, Roeske T, Syed Z, de Bruyne M, Galizia CG (2006) The molecular receptive range of an olfactory receptor *in vivo* (*Drosophila melanogaster* Or22a). J Neurobiol 66:1544–1563

82. Sachse S, Rueckert E, Keller A, Okada R, Tanaka NK, Ito K, Vosshall LB (2007) Activity-dependent plasticity in an olfactory circuit. Neuron 56:838–850

83. Silbering AF, Galizia CG (2007) Processing of odor mixtures in the *Drosophila* antennal lobe reveals both global inhibition and glomerulus-specific interactions. J Neurosci 27:11966–11977

84. Sachse S, Galizia CG (2003) The coding of odour-intensity in the honeybee antennal lobe: local computation optimizes odour representation. Eur J Neurosci 18:2119–2132

85. Olsson SB, Kuebler LS, Veit D, Steck K, Schmidt A, Knaden M, Hansson BS (2011) A novel multicomponent stimulus device for use in olfactory experiments. J Neurosci Methods 195:1–9

86. Vetter RS, Sage AE, Justus KA, Cardé RT, Galizia CG (2006) Temporal integrity of an airborne odor stimulus is greatly affected by physical aspects of the odor delivery system. Chem Senses 31:59–369

87. Galizia CG, Vetter RS (2004) Optical methods for analyzing odor-evoked activity in the insect brain. In: Christensen TA (ed) Advances in insect sensory neuroscience. CRC Press, Boca Raton, pp 349–392

88. Laissue PP, Reiter C, Hiesinger PR, Halter S, Fischbach KF, Stocker RF (1999) Three-dimensional reconstruction of the antennal lobe in *Drosophila melanogaster*. J Comp Neurol 405:543–552

89. Truong K, Sawano A, Mizuno H, Hama H, Tong KI, Mal TK, Miyawaki A, Ikura M (2001) FRET-based in vivo Ca^{2+} imaging by a new calmodulin-GFP fusion molecule. Nat Struct Mol Biol 8:1069–1073

90. Mank M, Reiff DF, Heim N, Friedrich MW, Borst A, Griesbeck O (2006) A FRET-based calcium biosensor with fast signal kinetics and high fluorescence change. Biophys J 90:1790–1796

91. Ohkura M, Matsuzaki M, Kasai H, Imoto K, Nakai J (2005) Genetically encoded bright

Ca^{2+} probe applicable for dynamic Ca^{2+} imaging of dendritic spines. Anal Chem 77:5861–5869

92. Nagai T, Yamada S, Tominaga T, Ichikawa M, Miyawaki A (2004) Expanded dynamic range of fluorescent indicators for Ca2+ by circularly permuted yellow fluorescent proteins. Proc Natl Acad Sci 101:10554–10559

93. Mank M, Santos AF, Direnberger S, Mrsic-Flogel TD, Hofer SB, Stein V, Hendel T, Reiff DF, Levelt C, Borst A, Bonhoeffer T, Hubener M, Griesbeck O (2008) A genetically encoded calcium indicator for chronic *in vivo* two-photon imaging. Nat Methods 5:805–811

94. Tian L, Hires SA, Mao T, Huber D, Chiappe ME, Chalasani SH, Petreanu L, Akerboom J, McKinney SA, Schreiter ER, Bargmann CI, Jayaraman V, Svoboda K, Looger LL (2009) Imaging neural activity in worms, flies and mice with improved GCaMP calcium indicators. Nat Methods 6:875–881

95. Denk W, Svoboda K (1997) Photon upmanship: why multiphoton imaging is more than a gimmick. Neuron 18:351–357

96. Diegelmann S, Fiala A, Leibold C, Spall T, Buchner E (2002) Transgenic flies expressing the fluorescence calcium sensor Cameleon 2.1 under UAS control. Genesis 34:95–98

97. Asahina K, Louis M, Piccinotti S, Vosshall L (2009) A circuit supporting concentration-invariant odor perceptiokn in Drosophila. J Biol 8:9

98. Baird GS, Zacharias DA, Tsien RY (1999) Circular permutation and receptor insertion within green fluorescent proteins. Proc Natl Acad Sci 96:11241–11246

99. Griesbeck O, Baird GS, Campbell RE, Zacharias DA, Tsien RY (2001) Reducing the environmental sensitivity of yellow fluorescent protein. J Biol Chem 276:29188–29194

100. Yu D, Baird GS, Tsien RY, Davis RL (2003) Detection of calcium transients in *Drosophila* mushroom body neurons with Camgaroo reporters. J Neurosci 23:64–72

101. Ma J, Ptashne M (1987) A new class of yeast transcriptional activators. Cell 51:113–119

102. Root CM, Masuyama K, Green DS, Enell LE, Nässel DR, Lee C-H, Wang JW (2008) A presynaptic gain control mechanism fine-tunes olfactory behavior. Neuron 59:311–321

Chapter 4

Functional Imaging of Antennal Lobe Neurons in Drosophila with Synapto-pHluorin

Dinghui Yu and Ronald L. Davis

Abstract

A method for imaging the synaptic activity of antennal lobe neurons in the *Drosophila* brain was developed to visualize and study cellular memory traces. Cellular memory traces are defined as any change in the activity of a neuron that is induced by learning, which subsequently alters the processing and response of the nervous system to the sensory information that is learned. Synapto-pHluorin (spH), a protein fusion of a pH-sensitive mutant of green fluorescent protein with synaptobrevin, was expressed in the projection neurons of flies. This protein, by virtue of its pH sensitivity, registers synaptic activity with increased fluorescence due to its movement from the lumen side of the synaptic vesicle to the synaptic cleft upon vesicle release. The steps required to measure synaptic activity using this fluorescent reporter in living flies in response to sensory cues and learning include: (McGuire et al., Prog in Neurobiol 76:328–347, 2005) microsurgery of transgenic flies to visualize spH fluorescence, (Davis, Annu Rev Neurosci 28:275–302, 2005) application of odor and electric shock cues to the fly being imaged, (Keene and Waddell, Nat Rev Neurosci 8:341–354, 2007) microscopic monitoring of the fluorescence before, during, and after the presentation of sensory cues, and [Miesenböck, Nature 394:192–195, 1998] analyzing the changes in fluorescence across time and space. One example of how this methodology has revealed insights into learning processes is discussed. The pairing of an odor (conditioned stimulus, CS) with electric shock pulses (unconditioned stimulus, US) produced a change in the synaptic response of PNs to the trained odor.

Key words: Optical imaging, Neural activity, Olfactory learning, Olfaction, Antennal lube

1. Introduction

Drosophila melanogaster has been used as a model system to study the molecular and circuit mechanisms of learning and memory for more than 30 years (1, 2). The ease and rapidity of genetic manipulation in *Drosophila* along with the rapid development of new tools and techniques by the *Drosophila* research community has made the fly an ideal system for studying a wide array of biological and medically-relevant issues. Many of the early studies focused on the molecular mechanisms of memory formation. For instance,

Jean-René Martin (ed.), *Genetically Encoded Functional Indicators*, Neuromethods, vol. 72,
DOI 10.1007/978-1-62703-014-4_4, © Springer Science+Business Media, LLC 2012

combined molecular genetic and behavioral studies demonstrated that memory formation depends critically on the cAMP-dependent protein kinase (PKA)-CREB signaling pathway that is conserved between vertebrates and invertebrates (for recent reviews see refs. (2, 3)). More recent investigations have employed a systems level approach to identify the circuitry and systems logic underlying memory formation, storage, and retrieval. Probing of memory circuits at the systems levels in many organisms have utilized direct electrophysiological recordings from brain neurons, but these are difficult to perform in living insects because of the small size of the neuronal cell bodies. In addition, electrophysiological readouts from the cell bodies may not detect or may misrepresent the important physiological events that occur within neuronal processes or at the synaptic junctions between neurons. Much traction, however, has been gained in studies of neural activity using functional optical imaging with fluorescent proteins like the many derivatives of green fluorescent protein (GFP).

Many different derivatives of GFP have been introduced as tools for studying biological processes. One derivative that is particularly important to neuroscience studies is synapto-pHluorin (spH) (4). This derivative is a protein fusion between a pH-sensitive mutant of GFP with synaptobrevin, constructed in a fashion so that the GFP moiety resides on the lumen side of the synaptic vesicle. The ecliptic version of spH exhibits modest fluorescence at pH environments of less than pH 6 under 470 nm excitation. Fluorescence increases within 20 ms after entering a pH environment that is close to neutral. Thus, when spH is targeted to synaptic vesicles that are located subsynaptically, it is relatively non-fluorescent because of the acidic environment of the vesicle. But it produces a burst in fluorescence as the vesicle fuses with the synaptic plasma membrane and spH becomes exposed to the neutral pH environment of the synaptic cleft.

Drosophila can form memory from many different types of events and tasks including classical conditioning (5). The most popular classical conditioning protocol produces memory by employing an association between and odor conditioned stimulus (CS+) and the unconditioned stimulus of electric shock (US). Flies display memory of this association in a test of their aversion of the CS+ odor relative to a second odor, the CS−, which is also presented during training but unpaired with the US. A major issue in neuroscience concerns the identity of the neurons that integrate the two stimuli, the CS+ and the US, in a way that leads to subsequent avoidance of the CS+. Potentially, these neurons may increase synaptic release in response to the CS+ after the animal has been trained. To identify these neurons and the nature of the memory traces that are formed from such training, we developed methods to optically measure synaptic release in flies that have been trained or pseudotrained.

The *Drosophila* olfactory system is comprised of the antenna, the antennal lobe (AL), the mushroom bodies (MBs) and the lateral horn (6). The pathway of the sequential transfer of olfactory information through the brain is from antenna, to the AL, and to the MBs and lateral horn. Olfactory receptor neurons (ORNs) in the antenna project their axons to glomeruli in the AL, the insect counterpart of vertebrate olfactory bulb. The glomeruli—morphologically identifiable regions of synapse-rich neuropil—consist of neurites and synaptic junctions from four types of neurons: the ORNs, inhibitory GABA-ergic local interneurons, excitatory local interneurons, and excitatory projection neurons (PNs). Reciprocal excitatory connections mediated by dendrodendritic cholinergic synapses and gap junctions exist between local excitatory interneurons and PNs (7). Output of olfactory information from the AL is the function of PNs. PNs send axons to MB calyx, the neuropil area containing synapses between PNs and MB neurons, and to the lateral horn. The MB neurons, in turn, project their axons into separate neuropil areas known as the lobes. There are five lobe regions, α, α', β, β', and γ, representing the axon branches of the three classes of MB neurons: α/β, α'/β', and γ.

Important tools in addition to the fluorescent reporters for optical imaging of neural activity are *Drosophila Gal4* lines. There exist tens of thousands of *Gal4* lines available in the *Drosophila* research community that allow expression of a separate transgene (*Uas-spH*, for instance), in virtually every type of neuron in the olfactory nervous system (8–11). One *Gal4* line in particular, *GH146-Gal4*, produces expression of *Uas*-transgenes in about 50 % of the AL PNs (12). With the expression of spH directed specifically to a subset of PNs in the AL, the optical imaging of synaptic activity in behavioral conditioned *Drosophila* brains became feasible. Individual flies are conditioned using several different procedures and the synaptic responses of PNs in the AL measured before and after conditioning to specific odors to help understand the nature of olfactory memory traces.

2. Materials

2.1. Transgenic Animals

The *Gal4/Uas* system was used to express reporter genes in the specific brain regions of *Drosophila* (13, 14). The reporter gene ecliptic spH (4) was inserted downstream of *Uas* sequences in the vector *pPBretU-H/X* (15). A total of five independent transformants containing ecliptic spH, *P{Uas-spH}*, were obtained and their insertions mapped genetically to different chromosomes. One transformant, *P{Uas-spH}34*, was mapped to the third chromosome and used for this study. *Gal4* lines contain the coding sequences for the yeast transcription factor, *Gal4*, at different sites in the genome. In most *Gal4* lines, the expression of *Gal4* in time and space is dictated by adjacent and anonymous genomic sequences. For instance, *GH146-Gal4* is a *Drosophila* line with *Gal4* expression in about half of the 180 PNs of the *Drosophila* AL (12). Progeny flies obtained by crossing *GH146-Gal4* flies with *P{Uas-spH}34* mates exhibit *spH* expression in these PNs. The *OR83b-Gal4* and *GAD-Gal4* lines express *Gal4* relatively specifically in ORNs and local inhibitory neurons (LNs), respectively (10). They are used in a similar fashion to obtain flies that express spH within their respective expression domains. Flies are maintained on a 12 h dark/12 h light cycle on standard *Drosophila* medium at 24°C.

2.2. Imaging

Although two-photon microscopes offer the ability to image at deeper levels than one-photon (confocal) microscopes, we found that the optical clarity of the brain is sufficiently clear to image all structures dorsal to the AL. Thus, the vast majority of imaging employs a standard confocal microscope.

2.3. Chemicals

Odorants used to deliver olfactory cues to flies include 3-Octanol (Oct, Aldrich) and 4-methylcyclohexanol (Mch, Sigma-Aldrich).

---

3. Methods

3.1. Microsurgery

Flies bigenic for the *Gal4* transgene and *Uas-spH* are collected after eclosion and kept on fresh media for 4 days before microsurgery. They are then coaxed into a 200 μl plastic pipette tip and forced to the narrow end with a small amount of air pressure. The antennae and dorsal head are exposed by trimming the pipette tip and the exposed head then secured to the tip opening with silicon cement (Kwik-Sil, from WPI). A region of cuticle approximately 200 μm by 200 μm is then removed from the top of the head using the tips of syringe needles. The window cut in the cuticle is subsequently covered with a piece of plastic wrap to prevent desiccation. The plastic wrap is sealed into place with a bead of silicon cement. Flies prepared in this manner remained viable for at least 2 h without obvious desiccation.

3.2. Functional Imaging

Functional imaging is performed on the prepared flies after mounting them beneath the 20× objective of a Leica TCS confocal microscope (Fig. 1). The spH molecules are excited with 488 nm laser light with the emissions collected at 520 ± 15 nm. Images are acquired at approximately five frames per second at a resolution of 256×256 pixels. Each imaging session lasts about 24 s.

The optical window in the dorsal cuticle allows the visualization of basal fluorescence in 8 glomeruli in *GH146-Gal4 > Uas-spH* flies.

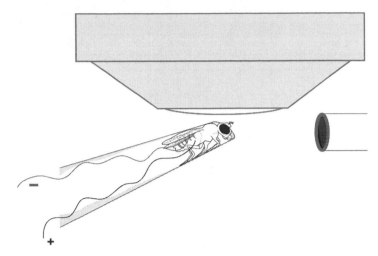

Fig. 1. The immobilized fly is mounted in a pipette tip under the objective of an upright confocal microscope for imaging. The odor is applied from the teflon valve system (Fig. 2) and through a puffer pipette to the antennae of the fly. Wires placed through the barrel of the pipette tip are positioned to deliver electric shock on command. Modified with permission from Davis (17).

These glomeruli include DM6, DM2, DM3, DL3, D, DL2, DA1, and VA1 (16). In addition to *spH* located at synaptic release sites within these glomeruli, some basal fluorescence also emanates from spH molecules located in the plasma membrane of the neurons (4). Two criteria are used to ensure that the x-by-y imaging is performed at the same z level between flies. First, all 8 glomeruli need to be visible and distinguishable in the basal fluorescent image prior to functional imaging. Large variations in the pitch and roll axes between fly preparations preclude the ability to make proper "between fly" comparisons. In such cases, the fly is discarded and another prepared. Second, the preparation is first scanned in the z-plane to locate the most intense focal plane, which occurs when the focus is centered on the midpoint of glomeruli in the dorsal/ventral axis. The fly is then functionally imaged in *xyt* after establishing these criteria.

Ten microliters of concentrated odorant is diluted with 100 µl of mineral oil. One microliter of this dilution is spread on a small piece of filter paper inside a syringe barrel that is placed in line with pressurized air flowing at a rate of 100 ml per min. Odorant delivery is accomplished using a 3-way teflon valve under the control of a programmable timer, such that fresh air is delivered to the animals for a determined period of time with an instantaneous switch to odor-laced air without altering the overall flow rate (Fig. 2). The odorant and air is directed at the antenna through a glass micropipette.

For establishing odor responses before and after conditioning, the odor is delivered through the micropipette for 3 s. Ten seconds

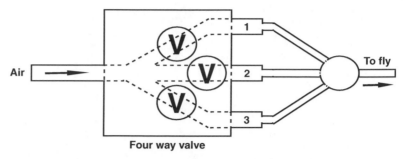

Four way valve

Fig. 2. Odorant delivery. The delivery of odorants is accomplished with a four-way teflon valve under the control of a programmable logic controller. The four-way valve is comprised of three, 2-way valves (V), each of which is either open or closed. Pressurized fresh air is delivered from the valve system to the fly immobilized under the microscope (Fig. 1) for a predetermined period. The controller then initiates an instantaneous switch to odor-laced air without altering the overall flow rate, followed by a return to fresh air. There are three outputs through the valve and to the puffer pipette that is positioned close to the fly. Output 1 is continuously on and delivers fresh air. Output 2 contains a filter paper with odorant and provides an odor source. Output 3 contains a filter paper with a second odorant in situations when two different odors are utilized, such as when using discriminative conditioning protocols. The airflow rate is 100 ml/min.

of baseline imaging precedes the odor application and the offset of odor application coincides with a return to delivery of fresh air. This allows the recording of the fluorescence of spH prior to odor delivery, during odor delivery, and after odor delivery. In some cases, two different odors need to be delivered to the fly being imaged, one of which is coupled with electric shock punishment. In such cases, the interval between applying the odors is 5 min. For conditioning under the microscope, odor application coincides with the presentation of electric shock pulses. In this instance, odor presentation occurs over a 1 min period.

Electric shock from a Grass S44 stimulator is delivered to flies by passing platinum wires carrying DC current through the barrel of the pipette. These wires touch the immobilized fly on the abdomen and/or legs. Electric shock pulses of 90 V last 1.25 s and are presented every 5 s during odor stimulation at training.

3.3. Data Analysis

The $\%\Delta F/F_o$ for each pixel within a region of interest is calculated across the time of the imaging scan using custom software. The raw images are first smoothed with a 7×7 Gaussian convolution filter. The value F_o is calculated for each pixel within the region of interest as the fluorescence averaged over five successive frames prior to odor application. The ΔF value is calculated for each pixel within the region of interest as the difference between the highest average intensity averaged over five successive frames during the 3 s of odor application and F_o. The values $\%\Delta F/F_o$ for eight glomeruli are calculated to have a measure of the synaptic response of the PNs to the odor.

4. Typical Results

4.1. PN Responses to Odors

Dissected and living flies are mounted under the microscope and imaged as described above. The average $\%\Delta F/F_o$ is calculated for each glomerulus for each fly. A representative time course for the $\%\Delta F/F_o$ to Oct for the DM6 glomerulus is shown in Fig. 3. The data for the glomeruli responses from a group of flies ($n > 6$) for each odor is then examined. The important parameters gathered from these data include: (1) Responses within any given glomerulus but between flies should exhibit a similar response profile. For instance, all flies should exhibit a response to Oct similar to the response shown in Fig. 3 for glomerulus DM6. (2) Each odor should have a unique response pattern across glomeruli. Not all glomeruli within an individual fly respond equally to a given odor; some glomeruli are inactive to certain odors while others are active. For instance, the glomeruli D, DL2, DA1, VA1 exhibit no detectable responses to Oct. This gives rise to a "glomerulus pattern" that may represent odor quality that is conserved across flies.

4.2. PN Responses to Electric Shock

Electric shock is delivered to the abdomen and legs of the imaged flies using 1.25 s electric shock pulses at 90 V DC, with one shock pulse every 5 s across a 1 min period as described above. Electric shock pulses produce a transient but pronounced increase in the $\%\Delta F/F_o$ with each pulse in all glomeruli, indicating that these pulses produce synaptic release from all PNs synapsing in the glomeruli. The periodicity in the change in $\%\Delta F/F_o$ as analyzed by Fourier transformation should match the periodicity in the delivery of shock pulses.

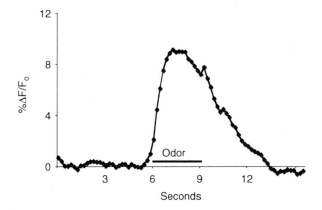

Fig. 3. Representative time course for the fluorescent response ($\%\Delta F/F_o$) to the odor Oct for the DM6 glomerulus. The response is calculated as the percent increase in fluorescence over baseline ($\%\Delta F/F_o$) as a function of time. For presenting the data in a more compact form (*bar graphs*), the $\%\Delta F/F_o$ is plotted as the percent difference between the maximum average intensity over five successive imaging frames during the 3 s odor application and the average intensity over five successive frames just prior to odor application.

**4.3. PN Responses
to Conditioning**

After the odor and the electric shock delivery are established and tested, the living flies can be subjected to conditioning protocols to determine whether this alters the subsequent response to the odor used for conditioning. Several different conditioning protocols can be used: (1) presentation of odor by itself (CS only), (2) presentation of electric shock by itself (US only), (3) presentation of odor and shock simultaneously (forward conditioning), (4) presentation of shock prior to odor (backward conditioning), or (5) presentation of shock after odor (trace conditioning). In all cases, these conditioning protocols are performed before (3 min before) and after (5 min later) an odor presentation by itself. This is to allow measuring the responses to odor before conditioning and then after conditioning, so as to gain an understanding of how each conditioning protocol might alter synaptic responses to odor.

Only forward conditioning alters subsequent responses to the trained odor. Specifically, forward conditioning with Oct recruits the D glomerulus into the representation of the odor and forward conditioning with Mch recruits the VA1 glomerulus (Fig. 4). In other words, the D glomerulus is not activated by Oct prior to conditioning and similarly, the VA1 glomerulus is not activated by Mch prior to conditioning. Forward conditioning alters the response patterns such that these glomeruli are activated by

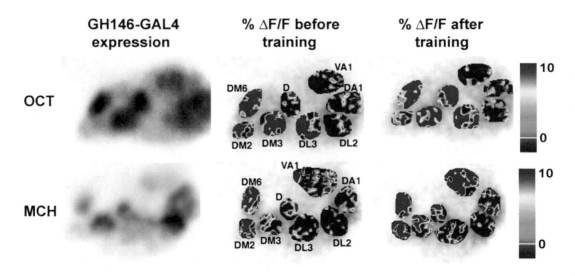

Fig. 4. Grayscale images of basal fluorescence of *GH146-Gal4/Uas-spH* are shown in the *left column* for Oct and Mch. Synaptic responses detected in the glomeruli due to odor stimulation before conditioning are illustrated in the *middle column* of images. Synaptic responses detected in the glomeruli due to odor stimulation after conditioning are illustrated in the *right column* of images. The change in fluorescence during odor application for each glomerulus is illustrated on a pixel-by-pixel basis as a pseudocolor image. PN synapses in four of the eight identifiable glomeruli exhibited synaptic responses to Oct before conditioning. After conditioning, PN synapses in glomerulus D also responded to the odor CS. PN synapses in three of the eight identifiable glomeruli exhibited synaptic responses to Mch before conditioning. After conditioning, PN synapses in glomerulus VA1 also responded to the odor CS (Adapted with permission from ref. (16)).

the respective odor. The duration of this recruitment into the glomerular representation of the odor is short lived, lasting only 5–7 min after forward conditioning.

The recruitment of PN synapses of the D and the VA1 glomeruli into the representation of Oct and Mch, respectively, after forward conditioning suggests that synaptic recruitment is odorant specific. Nevertheless, if the experimental animals are conditioned and tested with only one of the two odors, one cannot exclude the possibility that any recruitment could be generalized, or extend, to another odor. Discriminative conditioning, a protocol in which two odors are utilized with only one paired with electric shock pulses, is employed to dissect these alternatives. The goal of this type of experiment is to determine whether the changes observed by pairing electric shock with one odor become generalized, or whether the changes remain specific to the paired odor.

The protocol for this experiment is very similar to that for forward conditioning, but each odor is presented before pairing and then each odor is presented after conditioning as well. The presentation of the odors both before and after conditioning is separated in time by 3 min. When Oct and Mch are presented during discriminative condition with only Oct being paired with electric shock, the recruitment of the D glomerulus is specific to Oct. That is, Mch presentation after conditioning fails to activate the D glomerulus. Similarly, when Mch is paired with electric shock during conditioning, the recruitment of VA1 is specific to Mch presented after conditioning. There is no generalization to the alternative odor that is presented. Thus, glomerulus recruitment is odorant specific.

The memory trace represented by the recruitment of PN synapses described above could, in principle, occur through intrinsic mechanisms in the PNs. Alternatively, it is possible that the increased activity could simply reflect an increase that has occurred in neurons presynaptic to the PNs. This possibility can be tested by similar experiments in the neurons known to be presynaptic to PNs to determine whether forward conditioning alters their response patterns.

The PNs in the AL receive synaptic inputs from ORNs and LNs. To test whether a memory trace forms in the ORN presynaptic partners of PNs, the spH is expressed in the ORNs using a different Gal4 line, OR83b-Gal4 and synaptic responses before and after conditioning is measured in these neurons. A GAD-Gal4 line is used to direct expression of Uas-spH in inhibitory LNs to address the same issue. Forward conditioning is without effect on the synaptic responses of these neurons. In addition, these neurons are unresponsive to electric shock pulses, so they probably would not be able to integrate both CS and US information. Thus, the memory traces represented by the recruitment of the D or VA1 glomeruli appear intrinsic to the PNs.

5. Notes

5.1. Age of Flies

The age of flies is an important factor for experimental success. Flies of similar age exhibit similar basal fluorescence and keeping the age constant helps reduce variability. In addition, the cuticle of very young flies (<1 day) is too soft to dissect rapidly and cleanly, while the hemolymph of older flies (>10 days) tends to accumulate opaque substances that make imaging of all glomeruli simultaneously more difficult. Flies of 4–5 days offer a good balance between these factors. The flies at this age also exhibit sufficient spH expression for imaging.

5.2. Microsurgery

Microsurgery is probably the most important determinant in obtaining high quality results. The cuticle needs to be removed so that it does not block light penetration or emission. A layer of trachea under the cuticle also needs to be removed for effective observation. The window cut in the cuticle to allow visualization needs to be sealed effectively in order to prevent desiccation and allow the fly to remain healthy while imaging occurs.

Although dissection can be performed with sharp forceps or other types of surgical instruments, the use of two syringe needles offers good control in removing the cuticle and the underlying trachea simultaneously without disturbing the brain. This immediately exposes the clear brain, which is then sealed with a small piece of plastic wrap slightly larger than the window in the cuticle. These procedures need to be performed expeditiously over the course of a few minutes so that hemolymph does not evaporate. With practice, the microsurgery along with sealing the cuticle can be completed within 1–2 min.

5.3. Movement Control

A nagging problem with all in vivo and live imaging procedures is the movement of the objects being imaged. Motion artifacts that are inherent to such procedures can be controlled and often eliminated in several different ways. The head of the fly being imaged needs to be properly mounted and cemented to the pipette tip. In addition, cementing the back of the head to the anterior thorax with a small amount of silicon cement helps reduce movement during imaging. Some movement results from hemodynamic flow and occasional minor twitches due to the contraction of head muscles. Applying movement correction algorithms to the confocal time series can usually control these minor movements. These algorithms are often part of commercially available software and freeware, including Image J.

Acknowledgments

Research in the author's laboratory has been supported by grants from the NIH (NS052351 and NS19904).

References

1. McGuire SE, Deshazer M, Davis RL (2005) Thirty years of olfactory learning and memory research in *Drosophila* melanogaster. Prog Neurobiol 76:328–347

2. Davis RL (2005) Olfactory memory formation in *Drosophila*: from molecular to systems neuroscience. Annu Rev Neurosci 28:275–302

3. Keene AC, Waddell S (2007) *Drosophila* olfactory memory: single genes to complex neural circuits. Nat Rev Neurosci 8:341–354

4. Miesenböck G (1998) Visualizing secretion and synaptic transmission with pH-sensitive green fluorescent proteins. Nature 394: 192–195

5. Roman G, Davis RL (2001) Molecular biology and anatomy of *Drosophila* olfactory associative learning. Bioessays 23:571–581

6. Davis RL (2004) Olfactory learning. Neuron 44:31–48

7. Huang J, Zhang W, Qiao W, Hu A, Zhang A (2010) Functional connectivity and selective odor responses of excitatory local interneurons in *Drosophila* antennal lobe. Neuron 67: 1021–1033

8. Jefferis GS, Potter CJ, Chan AM, Marin EC, Rohlfing T, Maurer CR Jr, Luo L (2007) Comprehensive maps of *Drosophila* higher olfactory centers: spatially segregated fruit and pheromone representation. Cell 128: 1187–1203

9. Shang Y, Claridge-Chang A, Sjulson L, Pypaert M, Miesenböck G (2007) Excitatory local circuits and their implications for olfactory processing in the fly antennal lobe. Cell 128: 601–612

10. Ng M, Roorda RD, Lima SQ, Zemelman BV, Morcillo P, Miesenböck G (2002) Transmission of olfactory information between three populations of neurons in the antennal lobe of the fly. Neuron 36:463–474

11. Wang JW, Wong AM, Flores J, Vosshall LB, Axel R (2003) Two-photon calcium imaging reveals an odor-evoked map of activity in the fly brain. Cell 112:271–282

12. Stocker RF, Heimbeck G, Gendre N, de Belle SJ (1997) Neuroblast ablation in *Drosophila* P[GAL4] lines reveals origins of olfactory interneurons. J Neurobiol 32:443–456

13. Brand AH, Perrimon N (1993) Targeted gene expression as a means of altering cell fates and generating dominant phenotypes. Development 118:401–415

14. McGuire SE, Roman G, Davis RL (2004) Gene expression systems in *Drosophila*: a synthesis of time and space. Trends Genet 20:384–391

15. Roman G, He J, Davis RL (1999) New series of *Drosophila* expression vectors suitable for behavioral rescue. Biotechniques 27:54–56

16. Yu D, Ponomarev A, Davis RL (2004) Altered representation of the spatial code for odors after olfactory classical conditioning. Neuron 42:437–449

17. Davis RL (2011) Traces in *drosophila* memory. Neuron 70:8–19

Chapter 5

Performing Electrophysiology and Two-Photon Calcium Imaging in the Adult Drosophila Central Brain During Walking Behavior

M. Eugenia Chiappe and Vivek Jayaraman

Abstract

Continuing improvements in genetically encoded calcium indicators (GECIs) make imaging an increasingly attractive method to observe neural activity in the *Drosophila* brain. Two-photon imaging with GECIs allows calcium signals to be monitored in the entire adult fly central brain in vivo. It has recently become possible to perform two-photon imaging and electrophysiology in behaving flies during tethered flight (Maimon et al., Nat Neurosci 13:393–399, 2010) and walking (Seelig et al., Nat Methods 7:535–540, 2010). Here we elaborate on methods to perform two-photon calcium imaging with GCaMP, and GFP-/GCaMP-guided whole-cell patch clamp and loose-patch electrophysiological recordings in a head-fixed walking fly. We discuss how loose-patch recording can be used simultaneously with two-photon imaging to relate GECI signals to spiking activity in the central brain. Finally, we present a protocol for the loading and use of synthetic calcium dyes in combination with neurons labeled with a spectrally separated fluorescent marker to perform higher temporal resolution recordings from the central brain.

Key words: Two-photon, Calcium imaging, Electrophysiology, Behavior, Walking, Calibration, Rhod2-AM

1. Background and Historical Overview

Drosophila melanogaster has long been a genetic model organism of choice, but it was only recently that physiologists succeeded in targeting its brain. Early efforts to perform electrophysiological recordings from the fly brain were unsuccessful because of the small size of its neurons and the almost invisible glial sheath surrounding them. The advent of genetically encoded indicators, however, allowed researchers to record neural activity (1) and synaptic activity (2) from genetically identified neurons without desheathing the brain or using an electrode. Improvements in sensitivity, signal-to-noise, and kinetics have increased the appeal of GECIs (3, 4),

Jean-René Martin (ed.), *Genetically Encoded Functional Indicators*, Neuromethods, vol. 72,
DOI 10.1007/978-1-62703-014-4_5, © Springer Science+Business Media, LLC 2012

although they cannot match whole-cell patch clamp electrophysiology in temporal resolution and sensitivity to subthreshold changes in membrane potential (5).

In this chapter we elaborate on recent technical advances that have made it possible to record neural activity with two-photon imaging of GECIs and whole-cell patch clamp electrophysiology in awake, behaving flies (6, 7) (Fig. 1). These methods allow, for the first time, the physiology of identified neurons in the *Drosophila* brain to be monitored simultaneously with tethered locomotion. In addition to enabling studies that permit experimenters to correlate neural activity to the awake fly's behavioral responses (7), these methods have revealed that even neurons relatively near the sensory periphery, such as lobula plate tangential cells of the optic lobe, respond differently to identical sensory input depending on the behavioral state of the fly (6, 8).

Two-photon calcium imaging has several advantages over imaging with a CCD camera using visible light to excite calcium indicators. Two-photon imaging allows even the deepest structures in the fly brain to be imaged, and localizes the excitation light, which minimizes photobleaching and allows specific brain structures and processes to be monitored at high spatial resolution (beyond that achieved by using Gal4-driven expression of an indicator). In addition, the infrared wavelengths used for two-photon imaging are unseen by the fly and thus do not affect its behavior, except at higher laser intensities (discussed in a later section). An exhaustive description of the basic setup for two-photon GECI imaging in behaving *Drosophila* can be found in the recent literature (7). The requirements for performing whole-cell patch clamp technique recordings from adult fly brain neurons have been described thoroughly in a recent book chapter (9). In the sections to follow, we focus on some of the smaller details of imaging and two-photon-imaging-targeted patch recordings that can increase experimental reliability and robustness.

Despite their promise, most GECIs are still limited in their ability to act as good proxies for electrical activity (10–12). In order to interpret GECI output correctly, it is important to know what calcium responses mean in terms of electrical responses of the neuron. We describe an experimental paradigm to permit in vivo testing of GECI performance in the adult *Drosophila* brain using simultaneous 2-photon targeted electrophysiology. Such experiments demonstrated that the first sensors, e.g., GCaMP1.3 (13), missed a significant percentage of neural activity and suggested that the absence of a signal when using such sensors cannot be interpreted as sparseness of neural activity. Synthetic indicators, such as Rhod2-AM offer an alternative option for higher resolution monitoring of population calcium responses, and we detail procedures for their use to record activity from identified neurons in *Drosophila*.

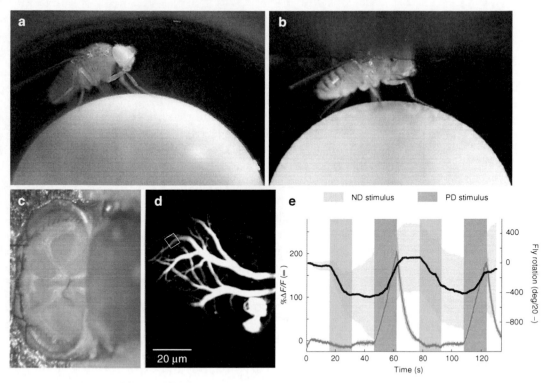

Fig 1. *Drosophila* head-fixed, walking-on-a-ball preparation. (**a**) High-speed video frame from a side view of a fly walking on a ball-like surface (a ballscape). We use such images as a reference for the natural ergonomics of walking on a 6 mm curvature. (**b**) Head-fixed, tethered fly walking on a ball from a side view. (**c**) Head-fixed, tethered fly from a dorsal view with the back of its head opened. Trachea and tracheal sacs are visible on top of brain tissue. (**d**) Collapsed, two-photon image stack from the lobula plate of the optic lobes of the fly showing two Horizontal System (HS) neurons, the HS-North and the HS-Equatorial. These neurons express GCaMP3.0 to record their responses to visual motion from different subcellular locations, for example as the one highlighted with the *yellow square*. (**e**) Simultaneous recording of optical signals from an HS-North neuron and walking activity of the fly when stimulated with wide-field gratings. HS neurons are direction selective neurons, i.e., their responses are selective to one direction of motion—the preferred direction (PD), which for this particular neuron is counterclockwise (*green traces*). The fly displays syndirectional walking activity: the fly follows the direction of the grating motion, a behavior known as an optomotor response (*black traces*).

2. Equipment, Materials, and Setup

2.1. Fly Genetics and Age

There is an ever-growing collection of *D. melanogaster* transgenic lines suitable for recording or perturbing the activity of a selected population of neurons. However, to analyze and correlate neuronal activity with behavior, it is important to assess the behavior under study in these transgenic stocks, which may differ from wild type strains. This is because the genetic background of transgenic flies, or the transgenes themselves, may affect the performance of a task. As a first step, we recommend evaluating the behavior of interest in transgenic flies and comparing it with that of wild type flies (Berlin, Canton-S, Oregon-R, etc.).

Using first generation (F1) flies from crosses between wild type and transgenic strains usually eliminates the effects of genetic background on behavior. However, when transgene expression is weak, homozygosing is a must. In these cases, the replacement of a chromosome with the wild type homologue may restore the wild type behavioral phenotype. This can be achieved either by several F1 backcrosses to parental wild type strains or by allelic recombination, the latter being a more tedious strategy but useful if the transgene is located at the chromosome that affects behavior.

The success of optical or electrical recording of neural activity in a behaving fly depends on several factors, notably the age of the fly. While dissections are easier with flies that have eclosed from the pupa less than 24 h from an experiment, protein levels (i.e., the expression of the transgene) tend to accumulate with time, making it easier to identify labeled neurons in older animals. This is, in fact, a nontrivial requirement. Moreover, certain behaviors demand a priming time, for example, adaptation to dewinged conditions, or isolation of males for aggression assays, making older flies more suitable to work with. In general, a good compromise between mature but not too old animals is 2–4 day-old flies. Thus, timing the stocks is a necessary step of fly husbandry for physiology.

Little is known about the origin of behavioral variability, but it can often confound the interpretation of experimental results. It is therefore important to keep rearing and experimental conditions, including temperature, humidity, and when the experiment is performed (in zeitgeber time, ZT), as reproducible as possible. It is well documented that circadian rhythms regulate fly's metabolism, physiology and behavior (14). We rear flies on standard cornmeal agar under a 12-h light and 12-h dark cycle at 25° and 45% RH; similarly, all experiments are performed at 25° and 45% RH during hours 10–14 ZT, when locomotion peaks (14).

2.2. Dissection Tools

Forceps: Dumonstar #55 further sharpened with sandstone (Dan's Whetstone Company, Inc)

Wax melter: Electra (Almore International Inc.)

Dissecting microscope: Leica M205C

Eyelash pencil: custom made

UV glue: Fotoplast Gel, Dreve, part number 44691

UV light: LED-100 UV portable, Electro-Lite Corp

CCD cameras: Teli or Cohu (2700 series) or Basler (e.g., 602f)

Video monitor: Triview, model TBM 1703

Lenses: Computar 25 mm

Cooling stage/plate: Custom made thermoelectrical controller to keep fly immobilized

2.3. Fly Holder

The fly holder is custom-made and the design will depend on the specifics of the space under the microscope. The major design constraints are:

- A rigid membrane to which the fly's head can be glued to limit brain movement during physiology. The membrane also needs to be thin (we use 0.025 mm thick stainless steel shim) to separate the legs, wings and sensory organs below from the extracellular saline that is used to bathe the brain above, which allows a water-immersion objective to be used.

- An opening in the membrane sized to accommodate the fly's head in a configuration that permits easy optical and electrical access to the neurons of interest.

- A "step" in the membrane behind the head to allow the rest of the fly's body (and/or wings) to be positioned naturally to permit walking or flying (Fig. 5.4b–d, "step").

For details of the construction of the holder we use, see Seelig et al. (7).

2.4. Behavioral Apparatus

Previous *Drosophila* walking experiments on an air-supported ball have used a light Styrofoam ball with a diameter of 7–9 mm. We use a ball of 40 mg and a diameter of 6 mm that allows robust optomotor behavior. Although a ball of lower mass can balance the fly's own inertia (mimicking natural walking in that respect), we find that flies easily lift balls of such lighter weights. A ball of larger diameter would provide a flatter surface for the fly to walk on, but would fill more of the fly's field of view. Our ball is manufactured out of polyurethane foam (Last-A-Foam, FR7120, General Plastics Manufacturing Company) using a bowl shaped file (7).

The ball's movements act as a proxy for the fly's walking activity. Ball velocity is tracked at high resolution using sensor chips from an optical mouse. The image of a small patch of the ball is projected onto optic flow detectors using appropriate lenses. We use two such cameras to track the ball and measure its velocity about three axes of rotation: pitch (forward rotation), roll (sideways rotation), and yaw (angular rotation). The ball's rotations can then be converted into forward, side and rotational velocities for the fly. Details of the assembly of the ball tracking system are found at http://www.flyfizz.org (7).

2.5. Two-Photon Imaging

Performing physiology on a fly walking on a ball requires a microscope with room underneath and around the objective. The ball and ball holder occupies the space that is usually occupied by a condenser and DIC optics in most commercial microscopes. Moreover, the two cameras tracking the ball are at 90° from each other and occupy a relatively large space. A custom-made, two-photon scope is perhaps the best solution for unconstrained space but many commercial microscopes may, with some modifications, also be used.

Since its original implementation (15), two-photon laser scanning microscopy has enabled recording transient calcium dynamics from distinct subcellular compartments of neurons deep in tissue (16–18). This is a particularly attractive technique when trying to understand how signals are processed in single neurons (19).

2.6. Electrophysiology: Electrodes, Solutions

Details about electrodes and internal solutions for whole-cell patch clamp recordings can be found in a recent review (9). In general, such electrodes have inner diameters of less than 0.5 μm at their tips, with impedances of approximately 6–10 MΩ. A desirable combination of small tip diameter and low impedance is achieved by using a pressure polishing technique first used by the worm (*Caenorhabditis elegans*) electrophysiology community (20). Electrodes used for loose-patch recordings typically have tips that are wider by a factor of 3 or even 4 than might be used for whole-cell patch clamp recording, and an impedance of 3–4 MΩ.

Loose-patch electrodes are filled with a solution containing 0.1 mM sulforhodamine B (Molecular Probes, Eugene, OR) in external saline solution. This allows pipette tips to be visualized with minimum power under the two-photon, and in a color channel easily separated from the green GECIs we typically use. The concentration of red dye in the pipette should be high enough to allow imaging at low laser power, but not so high that the dye clumps and affects the quality of the recording.

3. Procedures

3.1. Dissection: Fixing the Proboscis

For optical recordings it is important to have an immobile brain, because motion artifacts produce false positives in the signal (although this can be ameliorated with the use of ratiometric indicators, e.g., (3)). For electrophysiology, the more stable the brain is, the longer the recordings last. Three steps are necessary to achieve brain stability: (1) fix the proboscis to avoid proboscis extension and suction, (2) fix the head to the holder (see below), and (3) cut Muscle 16 to interrupt its pulsation (see below).

A single fly is anesthetized on ice and placed under a dissection microscope on a cold peltier stage to limit movement. The fly is moved to the center of the field of view with an eyelash pencil, and positioned on its back (Fig. 2a). A fine bended tungsten wire mounted on a 3D translation stage (Thorlabs) is gently positioned on top of the fly's forelegs (Fig. 2b). This prevents the legs from accidentally being waxed during the procedures to follow. Next, the proboscis is clamped at a fixed position with a low-melting-point wax mixture (1:1 molten bee wax and colophony, Sigma

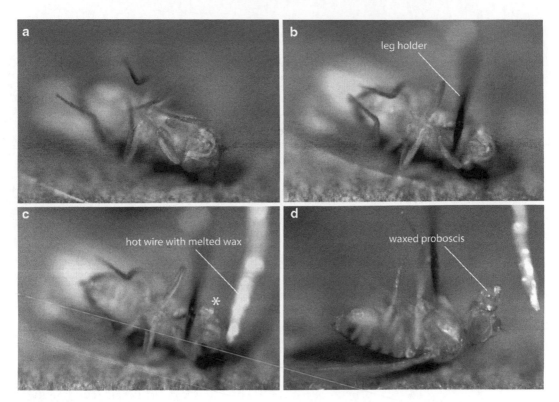

Fig. 2. Waxing the proboscis of the fly. (**a**) A cold anesthetized fly lies upside-down on a cool peltier platform. (**b**) A piece of bended insect pin is used to hold front legs and clamp the fly in place. (**c**) A hot wire that is attached to the wax melter approaches the dorsal hinge of the clypeous, inducing proboscis extension due to the heat (*asterisk*). (**d**) The proboscis is quickly bent and waxed.

Aldrich). The tip of a wax melter probe[1] (set at medium heat and dipped in wax mix) is advanced to the dorsal hinge of the proboscis rostrum, the clypeous (21) (Fig. 2c). The fly reacts to the high temperature of the probe by fully extending its proboscis. At this point its labellum is quickly pushed over the labrum to bend the midproboscis over and the labellum is glued to the clypeous (Fig. 2d). It is very important to perform this procedure as quickly as possible while the fly's proboscis is fully stretched. This is the only way to prevent cibarial pumping (proboscis suction), which produces brain motion when the waxed proboscis is only stretched midway. Finally, the underside of the stretched and bent proboscis is covered with wax.

Fixing the proboscis in the proper position is the first key step to achieve a good fly preparation for simultaneous physiology and behavior, especially for experiments aimed at recording optical

---

[1]Note that the probe from the wax melter is too large and needs to be polished to achieve a smaller tip. Use a file to achieve a convenient size between 0.3 and 0.5 mm. Alternatively, wrap the probe with a heat conducting wire to deposit wax on the fly's proboscis with a finer probe.

activity of neurons, or selectively labeling neurons with optical techniques such as photoactivatable GFP (PA-GFP). It is not unusual to have to fix the proboscis of a few flies before getting the perfect proboscis fixation.

3.2. Tethering the Fly

The goal of tethering a fly is to position it in the holder in a manner that permits well-controlled and repeatable experiments. Consistent alignment of the fly's head and body allows proper sensory stimulation and locomotion on the ball. Under the dissection scope, the fly is transferred to a metal chamber that is cooled by contact with the peltier stage (Fig. 3a). The chamber allows the fly to be set in a proper standing or hovering (22) position; a camera in front of the chamber permits alignment of the fly's yaw and roll body axis (Fig. 3a). A 5 mm long, 0.1 mm diameter tungsten rod positioned at a 45–60° angle with respect to the thorax of the fly is advanced with a 3D translator stage (Thorlabs) and glued to the fly between its head and notum using UV-activated glue.

3.3. Mounting the Fly into the Holder

The fly's head is positioned to permit optical and electrode access to brain regions of interest. Here we describe an orientation of the head that is suitable for electrical recordings from neurons located at the posterior part of the brain. Note that with two-photon imaging, anterior neurons can also be optically imaged in the same configuration. To access neurons located more laterally, e.g., those in the lamina, it may be necessary to slightly tilt the head to one side.

The tethered fly is inserted into a mount attached to a 3D micromanipulator to position the fly in the holder. The mount and micromanipulators are attached to a plastic platform to which the holder is clamped with screws or magnetic bases (Thorlabs) (Fig. 3b). The fly is carefully positioned in the center of the holder, just below the opening where the head is to be inserted (Fig. 4a, b).

The whole plastic platform is then mounted on a Z-translation stage to level the holder above the peltier plate. The fly's thorax and head are mounted on the cold stage surface to minimize the fly's movement. Two cameras orthogonal to each other show frontal and side views of the fly and the holder. These cameras and the microscope are used to align the head and body in the holder (Fig. 4c). The fly is then advanced just ahead of the edge of the opening. The fly is lifted with the micromanipulator and the height of the fly from the cold plate is adjusted with the Z-translator. Because the fly is placed slightly ahead of the opening, the dorsal part of the head is stopped by the holder platform and the head rotates forward through the "neck hinge" as the fly is lifted further (Fig. 4d). The fly is moved backwards to insert the bent head into the opening in the holder. The head may tilt about the yaw or roll axes, but this is easy to detect from the eyepieces of the microscope and the cameras at the front and side of the fly. The position of the head is adjusted with a clean, pulled glass capillary with a small

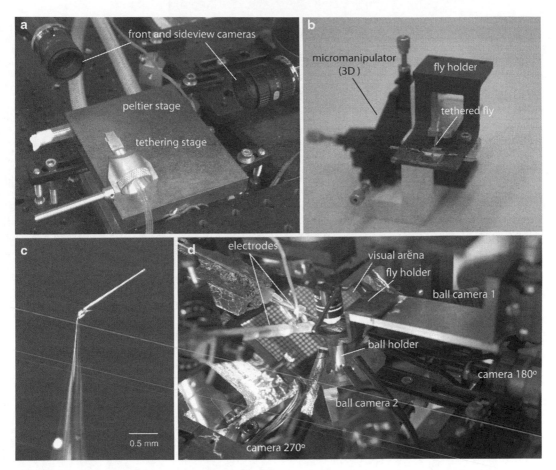

Fig. 3. Tethering and recording apparatus. (**a**) Tethering, waxing and dissection station. A metal plate running under a peltier cooling system is the platform for fly manipulation. Two cameras positioned orthogonal to each other provide frontal and side views for mounting the fly on the holder. Also shown is a metal tethering stage. (**b**) Fly holder and holder mount with 3-axis micromanipulator. Highlighted is a tethered fly that is introduced into the holder for mounting. (**c**) Glass capillary used to spread glue and fix the fly on the holder. (**d**) A head-fixed fly (Fig. 1b) in the fly holder with an exposed brain is mounted under the microscope and ready for visually stimulation (visual arena). Neural activity can be recorded by electrophysiological (electrodes) and optical (two-photon microscope, out of view) techniques. Ball camera 1 and 2 track the rotation of the ball as a proxy of the fly's responses to visual stimuli. 270°- and 180°-cameras are used to mount the fly on the ball properly.

glass ball at its end, and by moving the fly slightly (Fig. 3c). The side view of the fly shows how much of the eye is exposed on the underside of the holder (Fig. 4d).

Once the head is well aligned, it is glued to the holder with UV-activated glue (Fotoplast gel, Dreve). The glue is spread around the boundaries between the head and holder with the bulky glass capillary, and cured with UV light (LED-100 UV portable; Electro-Lite Corp) for 20s. Spreading a small amount of glue in small, selected areas, and quickly curing the glue with UV light avoids diffusion of the glue into the eyes. After gluing, the fly is lowered slightly and moved backwards with the micromanipulators

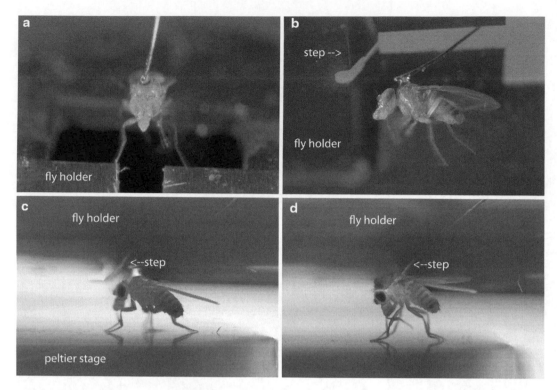

Fig. 4. Mounting the tethered fly on the holder with micromanipulators. (**a**) Frontal view of a tethered fly inserted in the open space of the holder. (**b**) The fly is lowered and introduced through a furrow to the head well. (**c**) The holder and fly are further lowered to have the fly in contact with the cool peltier to minimize her movements during mounting and alignment into the holder. (**d**) The fly's head is inserted into the head-well and aligned using the dissecting microscope and two cameras.

to relax neck tension and provide optimal optical access to the back of the head (Fig. 1c). The side view camera is used to ensure that the thorax and legs' positions are suitable for proper locomotion. The notum is glued to the holder following a procedure similar to that described for the head.

After the fly is firmly fixed, the holder's remaining opening is closed with a square piece of aluminum foil spread with vacuum grease. This avoids future liquid leakage. Importantly, any remaining toxic uncured glue is removed by using tissue paper, and by extensive rinsing with saline solution.

3.4. Dissecting the Brain of the Fly

Areal trachea and a membranous sheath rich in collagen covers the *Drosophila* brain. Trachea can significantly reduce optical access to some regions of interest (ROIs), and the sheath prevents seals from being made on somata during patch recordings. Therefore, depending on whether the experiment consists of optical imaging alone or electrophysiology, the steps of the dissecting procedure change slightly. In all cases, dissections are performed with the plate heated to room temperature to bring the fly out of cold-anesthesia. With the fly awake (and perhaps trying to escape), we are able to monitor how well the proboscis was fixed and, at the end of the dissection, whether or not Muscle 16 has been rendered inoperative (see below).

Under oxygenated saline solution (which is periodically changed during the dissection), the cuticle is removed with sharp forceps (Fig. 1c and (9)). The cuticle is first ruptured by gently touching it with the tip of the forceps. From that opening, and by pushing with the forceps' blade, the cuticle is removed following a contour just above the ROI. A broken patch pipette is used to clean fatty cells by suction. At this point, any brain pulsation due to Muscle 16's activity or improper proboscis fixation becomes apparent.[2] For optical imaging experiments, if the proboscis has not been fixed in the proper position, cibarial pumping occurs, and the fly must then be discarded. For electrophysiology, if brain movements due to cibarial pumping are not large, stable recordings can still be achieved, provided the stiffness of the patch electrode is reduced (for example, by elongating the taper to make the electrode mechanically more compliant).

Any trachea or tracheal sac over the ROI is gently removed with forceps. Note that the next step, which is to detach Muscle 16, critically depends on: (1) whether the experiment is designed for optical recordings and (2) if the experiment is designed for electrophysiology only. For some situations, disrupting Muscle 16's activity is not crucial; if that is the case, this step is avoided to minimize mechanical perturbation during the dissection. When the head is bent forward, the esophagus lies dorsal to the neck attachments, under an area of the cuticle that is darker and thicker. This area of the cuticle is removed by following the boundaries between the darker and lighter cuticle, and the fatty cells are removed as explained before. The brain shows periodic pulsations due to the Muscle 16 attachment (21). Muscle 16 traverses the brain along the anteroposterior axis dorsal to the esophagus. Its anterior attachment is located at the base of the antennal cup, and its posterior attachment is located at the posterior cuticle, next to the aortic funnel. Very carefully clipping and pulling the muscle at the dorsal aspect of the esophagus with sharp forceps disconnects the paired muscle from its posterior attachment. The muscles are difficult to see. However, when clipped and pulled, two thin filaments are snapped and the beating of the brain stops.

3.5. Mounting the Fly on the Rig

The head-fixed, tethered fly with the brain exposed is mounted under a custom-built two-photon microscope on top of an air-supported high-density foam ball with the help of a pair of orthogonally positioned cameras (7) and a 3D micromanipulator and goniometer attached to the holder's mount (Fig. 1b). The cameras,

---

[2] In general, the difference between brain motion due to muscle 16 activity vs. cibarial pumping is that the former is periodic while the latter occurs in bursting mode, with faster pulsations.

located at 180° and 270° from the fly's orientation, are focused on the fly, and provide posterior and side views that aid in positioning the fly so that it is centered on the north pole of the ball (Fig. 3d). The fly needs to be positioned at a height from the ball that readily induces walking and its free-walking posture is used as a guide for this.

3.6. Physiology in the Behaving Fly Brain

The area of interest to image or to target for whole-cell patch-clamp recordings is visualized with either a 40× or 60× water-immersion objective. If the ROI is optically accessible and there is no brain motion, the holder is connected to the perfusion system and the solution is kept oxygenated and at 21–24°C for the entire experiment. Otherwise, the preparation is brought back to the dissecting scope and the muscle-clipping step is repeated. For experiments designed for two-photon imaging, the ROI is located and the parameters of imaging are set up: scan rate, zoom, pixel resolution, wavelength, and laser power. We recommend imaging at a power that is less that 20 mW at the back aperture of the objective to avoid behavioral (startle) responses during scanning.

For electrophysiology, it is necessary to clean the collagen-rich sheath that covers the brain. If the transgenic line in use is specific, meaning that labeled cells are sparse and expressing fluorescent proteins such as GFP readily identifies target cells, it is possible to optically target and record from the cell somas. The target cells are identified under epifluorescence illumination or by using two-photon imaging. Using oblique infrared (IR) illumination (see below), the sheath is ruptured by the local application of a solution of collagenase (0.5–2 mg/ml) (6). A medium-sized patch pipette (tip ~2 μm) is filled with the collagenase solution and brought over the area where the target cells are. The bath temperature is then set to 29–30°C and the pipette is gently pressed on to the sheath while using positive pressure until a rupture of the membrane is observed (6). The temperature of the bath is then brought to 20°C to minimize further degradation of the tissue. With a cleaning pipette (tip ~5 μm), the exposed perineuroglia is then removed to access target cells.

If the transgenic line is not specific and target cells are difficult to optically isolate (Fig. 5a), they can still be specifically labeled and identified using photoactivatable or photoswitchable proteins (e.g., PA-GFP) in combination with 2-photon microscopy (23). To isolate neurons of interest, the two-photon beam is localized to a neuropile region where neurons of interest are known to project. Photoconversion is carried out by defining an ROI in the neuropile and illuminating the ROI with 710–810 nm light (24, 25). Photoconversion generally requires several iterations of excitation interleaved by resting periods to minimize photobleach and allow photoconverted PA-GFP molecules to diffuse along the neuron. We usually target the ROI with 810 nm of light for 10–20′ (256×256 pixel resolution, scanning at 2 ms/line, typically 20

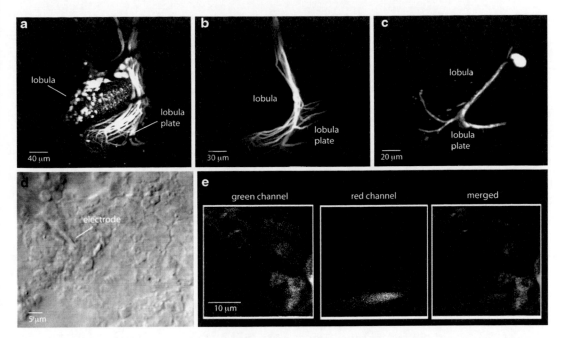

Fig. 5. Indentifying and recording from neurons in the *Drosophila* brain. (**a**) Neurons labeled by driving expression of GFP under the enhancer trap line *DB331*. Both the lobula and the lobula plate display green labeled neurons. Two-photon stacks. (**b**) If PA-GFP is expressed under the same enhancer-trap line, by photostimulating an area of the lobula plate where dendrites of lobula-plate tangential neurons reside, a group of vertical system (VS) neurons is specifically highlighted. (**c**) Furthermore, if photostimulation is further localized to a single cell soma then a unitary VS neuron is revealed. (**d**) Field of view of the dorsal cortex of the *Drosophila* brain illuminated by oblique infrared light. Note that the phase contrast permits visualization of cell somata and electrode's tip. (**e**) Two-photon guided patching. Target neurons are approached using two-channel imaging on the two-photon microscope.

averages of the ROI, repeated 10–30 times at 30–60″ intervals) (Fig. 5b). To visualize and define ROIs for photoconversion, we sometimes coexpress PA-GFP with a red fluorescent protein, or, alternately, scan with 925 nm light at high power (there is spurious PA-GFP activation in this condition). It is important to note that labeling of cells depends exclusively on the localization of the photoconversion beam. For example, a moving brain would induce a volume of photoconversion that is larger than the desired one. Furthermore, stability of the brain is crucial for neurons that do not project to a compact neuropile but have rather sparse terminals. In such cases, it is convenient to start photoconversion within segments of a single neuron's arborization and then finalize labeling at the target neuron's soma (Fig. 5c).

3.6.1. Procedures for Two-Photon Guided Patching

For general details about electrophysiology and the whole-cell patch clamp method, we refer readers to existing sources (26, 27). A recently published and excellent book chapter (9) provides detailed protocols for differential-interference-contrast-guided whole-cell patch clamp recordings from small *Drosophila* neurons. In this section, we focus only on the somewhat different procedures involved in targeting neurons using oblique illumination and two-photon imaging of GFP, PA-GFP or a GECI in a behaving preparation. We discuss both whole-cell and loose-patch recordings. The latter can be used to establish the extent of correlation between the spiking activity of a neuron and its GECI response.

One of the most important requirements for consistently successful whole-cell and loose-patch recordings is being able to clearly visualize the tip of the electrode and the membrane of the targeted cell. A walking or flying preparation has the ball with its holder or IR cameras in place of the conventional differential interference contrast (DIC) optic pathway of the microscope, thus preventing IR-DIC illumination. As an alternative, epifluorescent light may be a possible illumination source, but it has two important drawbacks. First, excitation of fluorescence causes phototoxicity, and, thus, the illumination time and light intensity need to be minimized. Second, the electrode is typically filled with a red fluorescent dye, and to visualize the tip of the electrode under fluorescence illumination simultaneously with the cell membrane, two independent cameras capturing different color channels are required. A better alternative is to use diffuse, oblique illumination and/or two-photon guided patching. Oblique illumination is easily obtained with a single IR-LED source placed at an oblique angle from the fly (Fig. 5d). To improve the contrast generated by oblique rays, a piece of black thin plastic or cardboard can be introduced at the LED as a patch stopper.

In two-photon guided patching, target neurons are approached using two-channel imaging (11, 28) on the two-photon microscope. Ideally, frame scans from the red and green channels should be merged online to produce a clear picture of the distance of the electrode tip from the cell body (Fig. 5e). It is helpful to scan only a small region as fast as still allows a clear definition of the electrode tip and cell body. This provides quick feedback as the electrode is moved close to the neuron of interest. Maintaining low positive pressure is important to prevent the glass electrode from picking up stray tissue/fat. This positive pressure is released as the electrode dimples the cell body (this should be visible with the oblique IR-illumination, but is difficult to see with the two-photon). This should result in the electrode forming a multi-giga-ohm seal with the neuron. For loose patch recordings, a technique that works well is to maintain very slight positive pressure to prevent a giga-ohm seal, and to then release it while pulling back the electrode to let the neuron cell settle into a low resistance (100–300 MΩ) seal that allows spikes to be reliably detectable for over 30 min.

3.6.2. Data Analysis for Loose Patch Electrophysiology: Spike Detection

To detect spikes in loose patch recordings, we use a combination of criteria that are kept constant for all traces recorded during an experiment. The current traces are first bandpass filtered (10 Hz–1 kHz) and then boxcar filtered (1 ms window). The filtered traces are passed through an amplitude threshold that is held constant for all trials of a given neuron. Negative peaks, which dominate the shape of the inverted action potential in an extracellular recording, are considered candidate spikes and can be subjected to further tests. Slope changes and template-based matching of the trace around the candidate spike are usually sufficient to resolve remaining ambiguities in all good quality recordings.

3.7. Using Two-Photon Imaging and Electrophysiology to Test GECI Performance

Correctly interpreting results of experiments with GECIs, particularly when GECI responses are used a proxy for neural activity, requires knowing how these responses correlate with electrical responses of neurons. This relationship is dependent on both neuron- and sensor-specific factors. Important performance metrics include sensor kinetics (how quickly the GECI changes fluorescence and decays back to baseline in response to a given calcium transient, the indicator's K_{on} and K_{off}), threshold (minimum change in local calcium levels required to produce a reliable change in fluorescence), signal-to-noise ratio (the change in fluorescence for different levels of electrical activity as compared to variation in the signal during spontaneous/baseline conditions), and toxicity (how much the sensor affects normal function of the neuron). Ideally, these tests would be performed with simultaneous whole-cell patch clamp electrophysiology from the specific neuronal process/soma being monitored with a GECI, permitting tight control and monitoring of local electrical activity. However, the whole-cell technique causes a rapid change in local GECI concentration because of dialysis, especially at the soma, limiting its use an accurate estimation of in vivo GECI performance in that compartment. For spiking neurons, there is however a different technique that can be used to monitor suprathreshold activity: loose-patch recording. This technique, and the related cell-attached recording technique, allows relatively noninvasive monitoring-albeit not control-of the neuron's spiking activity, without disrupting its internal contents.

3.8. Relating Two-Photon Calcium Imaging and Electrophysiology: Typical Results

Although loose-patch recordings do not permit control of the patched neuron, simply correlating recorded spiking activity with calcium signals can be useful (Fig. 6a–d). They can provide an estimate of how much neural activity the calcium indicator typically misses in the neurons of interest, and, if there is a consistent relationship between electrical activity and calcium signal, how much is actually occurring. This type of analysis is shown, for example, in the cases of GCaMP1.3-expressing projection neurons (PNs) of the antennal lobe (Fig. 6a–d). In this case, the sensor was found to have very slow kinetics, and a high threshold of activation, missing even sustained 40 Hz spiking from the PNs. However, in the regime that it was active, GCaMP1.3 was found to have a linear relationship with spiking (Fig. 6d).

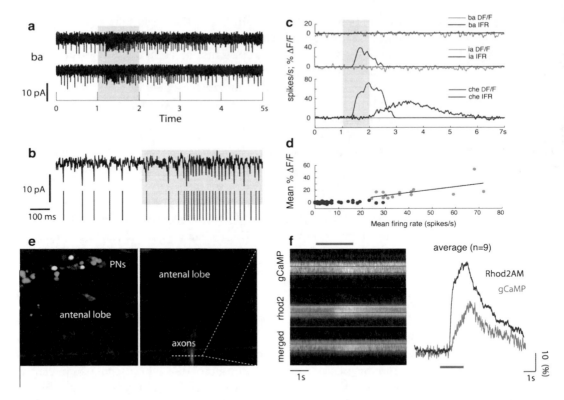

Fig. 6. Simultaneous electrophysiology and imaging in the *Drosophila* brain. (**a**) Olfactory projection neurons (PNs) were recorded under loose-patch (extracellular) recordings in response to an odor pulse (ba = benzaldehyde, ia = isoamyl acetate, che = cherry). (**b**) A template and threshold-based algorithm is used to isolate spiking activity highlighted by the *red ticks*. (**c**) Comparison of firing rates with optical signals obtained with GCaMP1.3 shows that calcium signals do not capture all electrical activity, and have a slow onset time. (**d**) Mean $\Delta F/F$ following odor presentation is partially correlated with the mean firing rate (during 1 s odor period) of various PNs in response to odor presentations. *Red dots* represent points that fall within 2.5 SDs of baseline variation. (**e**) Bulk loading of rhod 2-AM in the antennal lobes of *Drosophila*. *Left,* Two-photon stack of PNs some of which are loaded with Rhod 2-AM (*red*) and some other are labeled with GCaMP1.6 (*green*). Note that three PNs have both Calcium indicators in their cytoplasm. *Right,* Two-photon stack PN's axons exiting the antennal lobe. Indicated is a punctuated line that references where the line scans were performed. (**f**) Line scans of axons loaded with rhod 2-AM and expressing GCaMP1.6 (*yellow*). *Left,* Single trial line scan (500 Hz) during a 2 s odor pulse. *Right,* Averaged, integrated signals within the region of interest (*purple lines*) are shown as a function of time (*n* = 9).

4. Using Synthetic Dyes for Better Kinetics

GECIs have become the sensor of choice in optical imaging experiments because their expression can be restricted to specific cell populations by using the powerful *Gal4-UAS* approach. However, they do not yet match synthetic dyes in temporal resolution and kinetics. AM-dyes can thus provide a different solution in cases where these performance characteristics are particularly important. They can also be combined with genetic expression of non-Ca-sensitive fluorescent proteins to identity at least a subset of the imaged neurons by colocalization (29).

4.1. Additional Materials/Setup

Bulk loading with commercial calcium dyes depends crucially on a sonicator, for only constant sonication prevents precipitation of the dye.

4.2. Procedures

The dye solution is prepared on ice just before loading, i.e., after mounting the fly on the holder and opening its cuticle, but before desheathing and mounting the fly on the rig. The following protocol is an adaptation of a protocol developed for imaging in zebrafish and rodents (29, 30).

50 µg of Rhod2AM (Invitrogen) is used to make a solution of 1.25 µg/ml final concentration. 4λ of DMSO:F127 solvent mix (Invitrogen) is then added and the mixture is thoroughly sonicated on ice-cold water for about 20–30′. 34λ of extracellular solution is then added to the mixture and vortexed thoroughly. In addition, adding 2λ of Alexa to the mixture helps to visualize diffusion of the solution from the pipette. The mix is filtered in a 45 µm filter-Eppendorf with a short spin, and sonication is continued in ice-cold water for another 30–45′, during which time the dissection is continued.

4.2.1. Dye Loading

After desheathing the brain, a very small volume (no more than 2λ) is loaded into the pipette (medium size, tip of about 2–4 µm). The pipette is always loaded with solution that is still in the sonicator, in ice-cold water. It is important to have the sheath disrupted above the ROI. Most neuropiles in insects are also covered by a perineural glia, but this is easily penetrated by the pipette. Using two-photon scanning, the pipette is approached towards the target area, while delivering 0.5 Hz short pressure pulses of 5–10 ms duration with a picospritzer. If the tip of the pipette is not clogged, then a single-point diffusion pattern can be visualized with Alexa. Once in the neuropile of interest, several 2-Hz short pulses of 5–10 ms duration are delivered for about 5–10′. This can be repeated until penetration of the dye into the neuropile structure is observed. The preparation is under continuous perfusion of oxygenated saline solution. If the dye penetrates the neuropile, neurons pick it up and, with time, cell somas will be labeled (Fig. 6e). Typically, the fly is left unperturbed for about 15–30′ to allow the dye to diffuse along the neuron (Fig. 6e, right).

4.3. Typical/ Anticipated Results

In test experiments, we have combined the expression of GCaMP1.6 with bulk loading of Rhod2-AM in the antennal lobe of the fly. Rhod2-AM incorporates in both GCaMP expressing and nonexpressing neurons, allowing kinetics and response profiles comparisons between both calcium indicators (Fig. 6f). When comparing the two signals in the same neuron, Rhod2-AM responded faster than GCaMP as predicted by their different kinetics ((Fig. 6f, right) (4, 31)). Thus, for comparison and reconstruction of firing rates with calcium imaging, commercially available indicators are still better than GECIs (4). Using bulk loading of Rhod2-AM in transgenic fly lines expressing GFP in selected neurons enables some of the dye-loaded neurons to be identified, which is useful for many in vivo applications. This may be a good strategy for those interested in reconstructing firing rates out of fast calcium signals (29). Moreover, it can be also implemented in other insect species, where GECIs are not yet a possibility.

5. Conclusions

In this chapter we have provided a detailed description of steps to perform physiology in the central nervous system of *Drosophila* while the fly is walking on a ball. This procedure would need to be modified depending on the specifics of the microscope in use and the specific behavior under investigation. When possible, we encourage combining two-photon laser scanning imaging with electrophysiology. Not only do the two techniques provide complementary information, but simultaneous electrophysiology can also help researchers better interpret and calibrate data from calcium imaging experiments. Finally, we have described a procedure to use commercially available calcium sensors, which could be used as an alternative to GECIs if there are no enhancer trap lines labeling target brain areas, or if greater temporal precision is particularly important.

References

1. Fiala A, Spall T, Diegelmann S, Eisermann B, Sachse S, Devaud JM, Buchner E, Galizia CG (2002) Genetically expressed cameleon in Drosophila melanogaster is used to visualize olfactory information in projection neurons. Curr Biol 12:1877–1884

2. Ng M, Roorda RD, Lima SQ, Zemelman BV, Morcillo P, Miesenbock G (2002) Transmission of olfactory information between three populations of neurons in the antennal lobe of the fly. Neuron 36:463–474

3. Mank M, Santos AF, Direnberger S, Mrsic-Flogel TD, Hofer SB, Stein V, Hendel T, Reiff DF, Levelt C, Borst A et al (2008) A genetically encoded calcium indicator for chronic in vivo two-photon imaging. Nat Methods 5:805–811

4. Tian L, Hires SA, Mao T, Huber D, Chiappe ME, Chalasani SH, Petreanu L, Akerboom J, McKinney SA, Schreiter ER et al (2009) Imaging neural activity in worms, flies and mice with improved GCaMP calcium indicators. Nat Methods 6:875–881

5. Wilson RI, Turner GC, Laurent G (2004) Transformation of olfactory representations in the Drosophila antennal lobe. Science 303:366–370

6. Maimon G, Straw AD, Dickinson MH (2010) Active flight increases the gain of visual motion processing in Drosophila. Nat Neurosci 13:393–399

7. Seelig JD, Chiappe ME, Lott GK, Dutta A, Osborne JE, Reiser MB, Jayaraman V (2010) Two-photon calcium imaging from head-fixed Drosophila during optomotor walking behavior. Nat Methods 7(7):535–540

8. Chiappe ME, Seelig JD, Reiser MB, Jayaraman V (2010) Walking modulates speed sensitivity in Drosophila motion vision. Curr Biol 20:1470–1475

9. Murphy M, Turner G (2010) Chapter 18: in vivo whole-cell recordings in the Drosophila

brain. In: Zhang B, Freeman MR, Waddell S (eds) Drosophila neurobiology: a laboratory manual. Cold Spring Harbor Laboratory Press, New York

10. Pologruto TA, Yasuda R, Svoboda K (2004) Monitoring neural activity and (Ca²⁺) with genetically encoded Ca²⁺ indicators. J Neurosci 24:9572–9579

11. Jayaraman V, Laurent G (2007) Evaluating a genetically encoded optical sensor of neural activity using electrophysiology in intact adult fruit flies. Front Neural Circ 1:3

12. Reiff DF, Ihring A, Guerrero G, Isacoff EY, Joesch M, Nakai J, Borst A (2005) In vivo performance of genetically encoded indicators of neural activity in flies. J Neurosci 25:4766–4778

13. Nakai J, Ohkura M, Imoto K (2001) A high signal-to-noise Ca(2+) probe composed of a single green fluorescent protein. Nat Biotechnol 19:137–141

14. Allada R, Chung BY (2010) Circadian organization of behavior and physiology in Drosophila. Annu Rev Physiol 72:605–624

15. Denk W, Strickler JH, Webb WW (1990) Two-photon laser scanning fluorescence microscopy. Science 248:73–76

16. Yuste R, Denk W (1995) Dendritic spines as basic functional units of neuronal integration. Nature 375:682–684

17. Svoboda K, Helmchen F, Denk W, Tank DW (1999) Spread of dendritic excitation in layer 2/3 pyramidal neurons in rat barrel cortex in vivo. Nat Neurosci 2:65–73

18. Euler T, Detwiler PB, Denk W (2002) Directionally selective calcium signals in dendrites of starburst amacrine cells. Nature 418:845–852

19. Reiff DF, Plett J, Mank M, Griesbeck O, Borst A (2010) Visualizing retinotopic half-wave rectified input to the motion detection circuitry of Drosophila. Nat Neurosci 13:973–978

20. Goodman MB, Lockery SR (2000) Pressure polishing: a method for re-shaping patch pipettes during fire polishing. J Neurosci Methods 100:13–15

21. Demerec M (1950) Biology of Drosophila. Cold Spring Harbor Press, New York

22. Lehmann FO, Dickinson MH (1997) The changes in power requirements and muscle efficiency during elevated force production in the fruit fly Drosophila melanogaster. J Exp Biol 200:1133–1143

23. Patterson GH, Lippincott-Schwartz J (2002) A photoactivatable GFP for selective photolabeling of proteins and cells. Science 297:1873–1877

24. Datta SR, Vasconcelos ML, Ruta V, Luo S, Wong A, Demir E, Flores J, Balonze K, Dickson BJ, Axel R (2008) The Drosophila pheromone cVA activates a sexually dimorphic neural circuit. Nature 452:473–477

25. Ruta V, Datta SR, Vasconcelos ML, Freeland J, Looger LL, Axel R (2010) A dimorphic pheromone circuit in Drosophila from sensory input to descending output. Nature 468:686–690

26. Sakmann B, Neher E (1995) Single-channel recording, 2nd edn. Plenum Press, New York

27. Molleman A (2003) Patch clamping. Wiley, West Sussex, England

28. Margrie TW, Meyer AH, Caputi A, Monyer H, Hasan MT, Schaefer AT, Denk W, Brecht M (2003) Targeted whole-cell recordings in the mammalian brain in vivo. Neuron 39:911–918

29. Yaksi E, Friedrich RW (2006) Reconstruction of firing rate changes across neuronal populations by temporally deconvolved Ca²⁺ imaging. Nat Methods 3:377–383

30. Sato TR, Gray NW, Mainen ZF, Svoboda K (2007) The functional microarchitecture of the mouse barrel cortex. PLoS Biol 5:e189

31. Mao T, O'Connor DH, Scheuss V, Nakai J, Svoboda K (2008) Characterization and subcellular targeting of GCaMP-type genetically-encoded calcium indicators. PLoS One 3:e1796

Chapter 6

In Vivo Optical Recording of Brain Interneuron Activities from a Drosophila Male on a Treadmill

Soh Kohatsu, Masayuki Koganezawa, and Daisuke Yamamoto

Abstract

Recent development of genetically encoded calcium indicators (GECIs) allows us to directly monitor neural activities in *in vivo* preparations from various organisms, enabling the exploration of neural substrates for complex behaviors. As a showcase for such renovated neuroethology with GECIs, we describe a novel method to record optically interneuron activities from the brain of a tethered male fruit fly *Drosophila melanogaster* when contact chemosensory stimulation is given to his foreleg. In this protocol, the recording is made on the preparation in which a fly is put on an air-supported styrofoam ball, keeping the "moving" fly in position to allow monitoring of the activity of central neurons from the fly whose movement is minimally restricted. This protocol is applicable to the analysis of other behaviors induced by stimuli of different modalities with small modifications.

Key words: Calcium imaging, *Drosophila*, Courtship, Yellow Cameleon, *Fruitless*

1. Introduction

A large collection of sophisticated genetic tools in *Drosophila melanogaster* provides neuroscientists with the means by which to manipulate the activity of a genetically identified subpopulation of neurons in a freely moving animal (1–6). By observing the behavioral consequences of artificial activation/inactivation of particular neurons with transgenes, one can determine which neuronal assembly is the key building block of the circuit underlying particular behavior, and this approach is also powerful in elucidating dynamic changes in the circuitry associated with learning and memory (7–11).

Complementing the neural mapping with artificial activation/inactivation tools, optophysiology with genetically encoded calcium indicators (GECIs) offers the means to record activities from manipulated neurons, providing the information necessary to

Jean-René Martin (ed.), *Genetically Encoded Functional Indicators*, Neuromethods, vol. 72,
DOI 10.1007/978-1-62703-014-4_6, © Springer Science+Business Media, LLC 2012

understand how the neural circuit works. Optophysiology with GECIs overwhelms electrophysiology in that it enables one to monitor the activity of a larger number of neurons in a less invasive manner, while sparing temporal resolution. In practice, optophysiology offers higher reproducibility in recording the activity from an identified population of neurons across different individuals, and thus improves the chance to search and establish the physiological correlates of behavior, which is typically difficult to induce in repetition under stringent experimental conditions (12). In addition, given the large collection of fly strains available for driving transgene expression and rapid development of optical probes and imaging devices such as the two-photons confocal microscope, the potential targets for optophysiology continue to expand; for example, neurons located in the position not accessible by the electrode are now subjected to optical recording (13–19).

In exploring the neural substrate for a behavior in the central nervous system, it is of primary importance to carry out physiological recordings under the conditions where the fly retains the motivation to perform the behavior. The treadmill system, in which a tethered fly walking on an air-suspended styrofoam ball is subjected to physiological recordings, is a powerful tool for this purpose and provides a stable platform for fine behavioral and physiological measurements (12, 20, 21). This system, originally invented to analyze simple behaviors, such as optomotor response (20), proved to be extremely useful for the study of complex behaviors such as courtship: a male on the treadmill follows a courtship target and generates love songs by vibrating his wings, two hallmarks of the initial behavioral sequences of courtship (22). In studying courtship behavior, the treadmill system is particularly advantageous because it makes the contact chemosensory organs on the legs accessible for stimulation, which is otherwise undeliverable onto radically moving legs, even though this sensory modality is vital for controlling the behavior (22). In the physiological measurement from central neurons, the treadmill system, compared with the strictly immobilized preparation, should yield activities closer to those occurring in freely moving animals. Indeed, an experiment with a similar treadmill has successfully revealed that, when flies are actively walking, motion-detector interneurons become tuned to higher image motion speeds than when they are stationary (12). It is conceivable that neural activities associated with courtship are similarly susceptible to walking-dependent modulation.

In the following protocol section, we describe how to record calcium responses in the arbors of brain interneurons from a male fly that is placed on a treadmill and given contact chemosensory stimuli, i.e., nonvolatile sex pheromones, on his foreleg to trigger courtship (22, 23). By targeting the GECI Yellow Cameleon2.1, with a genetic mosaic technique, to a small number of cells, one can reliably identify the neurons from which activity recordings were obtained. The P1 cluster, a male-specific subpopulation of *fruitless*-expressing interneurons that have been implicated for triggering courtship, was thus shown to be activated by the stimulation with CHCs applied on the male foreleg (22).

2. Materials and Equipment Setup

2.1. Fly

- Subject fly: Flies carrying appropriate Gal4 and UAS transgenes to target GECI expression to the neurons of interest. As an example, here we use the Gal4 enhancer trap line *fru^{NP21}* with a Gal4 insertion in the *fru* locus together with a UAS line to express Yellow Cameleon2.1 obtained from a public stock center (Bloomington Drosophila Stock Center, http://flystocks.bio.indiana.edu/) (24). Isolate the males within 6 h after eclosion and keep singly in a vial for 5–7 days until used for the experiment.

- Stimulant fly: Females or males of the Canton-S wild-type strain. Collect the flies within 6 h after eclosion and keep in vials as groups of approximately ten individuals of the same sex for 5–7 days.

2.2. Fly Preparation

- Fly holder: Punch a hole approximately 8 mm in diameter on a plastic coverslip and cover it with a piece of polyethylene wrap using UV glue (Fig. 1a).

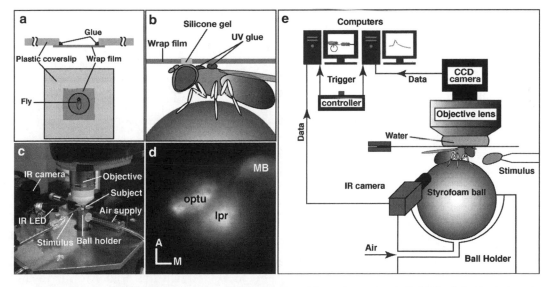

Fig. 1. Preparation and experimental setup for in vivo Ca²⁺ imaging with a male on a treadmill. (a) The fly holder for imaging neural activities. Top *Lateral view*, Bottom *Horizontal view*. (b) A schematic illustration of a male fly fixed to the holder for recording. *Lateral view*. (c) Arrangement of the experimental setup including the stimulus fly's body and the instruments for recording. (d) A raw YFP fluorescent image acquired through the window made on the head cuticle. *Horizontal view*. *A* anterior, *M* medial, *lpr* lateral protocerebrum, *optu* optic tubercle, *MB* mushroom body. (e) A diagram of the recording system.

- Syringe needle (24–27 gauge).
- UV curing glue and UV pen light.
- Silicone elastomer (Kwik-Sil: World Precision Instruments).
- Tungsten needle 0.15 mm in diameter attached to a holder.
- Fly Ringer's solution (NaCl: 130 mM, KCl: 5 mM, $MgCl_2$: 2 mM, $CaCl_2$: 2 mM, Sucrose: 36 mM, HEPES-NaOH (pH 7.3): 5 mM) (25).
- Cool anesthetization stage: A brass plate of $50 \times 50 \times 5$ mm.
- Pulled glass capillary.

2.3. Recording Setup

The recording system consists of two parts; one for acquiring fluorescent images of the brain and the other for monitoring behavior by the video recording of lateral views of the subject male (Fig. 1c, e). These two lines of recording system are synchronized by an electrical trigger signal. The subject fly is stimulated by touching his foreleg with a stimulant fly's body attached to a holder mounted on a manual micromanipulator.

- Standard upright fluorescent microscope.
- Water immersion objective lens with high NA values and a long working distance (W Plan-apochromat, x40, NA = 1.0, CarlZeiss).
- CCD camera and controlling software (C7780-20 and Aquacosmos, Hamamatsu photonics).
- High-pressure mercury lamp (HBO-100, CarlZeiss).
- Filter set: ND filters, 440/20 band-path filter, 455 dichroic filter and 460 nm long-path filter are used for preparation of excitation light and separation of emission light. Emission light is further divided with a beam splitting prism embedded in the CCD camera (460–490 nm for CFP and 490–570 nm for YFP channels), then projected on separate CCD chips.
- Infrared CCD camera (STC-H400, Sensor Technology Co. Ltd.).
- Macro zoom lens (12× zoom system, Navitar).
- Analog video timer (VTG-55D, FOR-A).
- A/D video converter (DVMC-DA2, Sony).
- Styrofoam ball (4–5 mm in diameter).
- Styrofoam ball holder: A brass base with a hemispherical pocket 6 mm in diameter and an air inlet 1 mm in diameter equipped on the bottom of the pocket. Attach this holder to the microscope stage with appropriate mechanics for fine adjustment of the position.
- Air source: Decompress the air stream from a compressor and pass through cotton, activated charcoal, and water in sequence

to make it clean, then introduce into the ball holder via a needle valve. Adjust the flow speed of the air so as to float the ball stably.

- Manual micromanipulator for handling the stimulus fly's body.

2.4. Data Analysis

- Software environment for image processing (ImageJ (26), http://rsbweb.nih.gov/ij/ and Excel, Microsoft).

3. Methods

3.1. Fly Preparation

1. Put a male in an ice-cooled glass vial up to 20 s for anesthesia.

2. On an ice-cooled preparation stage, fix the head to the thorax by filling the space between these structures with UV glue.

3. Put a small drop of UV glue on the dorsal tip of the thorax and fix it to the polyethylene wrap on the fly holder. The dorsal head region between ocelli and the base of the antennae must tightly attach to the wrap film (Fig. 1b). Prevent the glue from invading into the space between the head and the wrap film. Make sure that the antennae, especially the aristae do not contact the wrap film.

4. Put the holder in a humid chamber placing the fly ventral side down and let the fly hold a styrofoam ball, so as to prevent him from flying. Repeat procedures 1–3 to obtain the appropriate number of fly preparations. We usually prepare 12 flies at a time.

5. Keep the flies in a humid chamber for about 2 h for recovery.

6. Cut a square window on the wrap film covering the head using the edge of a syringe needle and put ~100 μl of Ringer's solution on the opening (Fig. 1b). Make sure that the Ringer's solution has not leaked out to the bottom of the wrap film. Sealing between the head and the cutting edge of the wrap film is not necessary at this stage, provided that the head is in tight contact to the wrap.

7. Cut a window on the head capsule with a syringe needle to allow visual inspection of the brain. Using fine forceps, remove fat bodies and gently unglue the air sacs from the brain and put them aside to expose the lateral protocerebral region (Fig. 1d). Take extreme care not to damage the air sacs.

8. Seal the head opening. Mix the two components of silicone elastomer. Remove the Ringer's solution covering the dorsal surface of the brain with micropipettes and a twist of kimwipe, then immediately apply a small drop of silicone elastomer to seal the head opening and to fill the space between the head

and the edge of the window on the wrap film. The silicone elastomer also serves to restrict the movement of the brain during recording. Leave it for 5 ~ min for curing.

9. Prepare stimulant flies. Remove the head and thorax from a cold-anesthetized fly by cutting the thoracicoabdominal junction using fine scissors. Insert a pulled glass capillary into the abdomen from anterior and seal the stump of the abdomen using UV glue.

3.2. Recording

1. Set the preparation on the microscope stage and look for raw YFP fluorescence originating from the lateral protocerebrum. Use minimal excitation light just sufficient for the detection of the fluorescent signal and detect it as quickly as possible to minimize the photo bleaching of YFP.

2. Adjust the position of the treadmill by monitoring the lateral view of the fly using an infrared camera. Make sure that the fly's legs of both sides evenly and gently come into contact with the styrofoam ball and he can easily walk on it.

3. Under dim red light, mount the stimulant fly's body on a holder and bring it with a micromanipulator close to the test fly's foreleg to be stimulated.

4. Decide the recording site (the region of interest (ROI)) for real-time monitoring of the fluorescent signal. Adjust the aperture of the excitation light not to illuminate the cuticular region around the head opening, so as to minimize the scattered light and improve the contrast. Set the ND filter, binning, and gain of the CCD camera so that the exposure of ~200 ms gives sufficient fluorescent signals in ROI (i.e., 1,000 ~ signal values in 12 bit camera) for both CFP and YFP channels.

5. Check the condition of the male. If he keeps running or shows active grooming, wait for ~5 min to calm him down. To apply a single stimulus by touching his foreleg, the male fly must be in the resting state. In case of a restless male, discard it and try another preparation.

6. Trigger recording. Each recording session could be ~20 s at the image acquisition rate of 5–10 Hz. Around 5 s after the onset of recording, move the stimulant fly's body forward to touch the test fly's foreleg, then quickly withdraw it by manually operating the micromanipulator, which is under the monitor through the infrared camera (Fig. 2a).

7. When successful, a rapid increase of the YFP signal and a decrease of CFP signal are observed just after the contact of the stimulant fly with the foreleg of the test fly, resulting in a rise in the YFP/CFP ratio (Fig. 2b). Under these conditions, a preparation typically yields records containing up to ten stimulation trials in an hour test session.

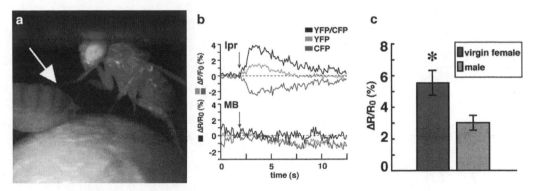

Fig. 2. Optical recording of calcium responses in *fru*-expressing neurons of a male as stimulated by a female contact with his foreleg (Modified from (22)). (**a**) A frame of video records showing the movement when the subject male is touching the stimulus female's abdomen with his foreleg. (**b**) Representative traces of calcium responses in the lateral protocerebrum and mushroom body. The touch stimuli were given at arrows. (**c**) A quantitative comparison of the response magnitude between the responses to a female body and those to a male body ($n = 10$). The mean ± SEM are shown. *$p < 0.05$ in the Mann–Whitney U test.

3.3. Data Processing

1. Determine the stimulus onset: Replay the video record captured by the infrared camera and determine the stimulus onset by picking the timestamp on the video frame where the foreleg of the subject male comes into contact, for the first time, with the stimulant fly's body.

2. Using ImageJ, register the stack of images to cancel drifts in the target position, if any, due to unavoidable movements of the brain.

3. Filter the images with a Gaussian filter (radius: 1–2).

4. Define the ROI, which is a circle of 10 pixels in diameter and also the background brain area not expressing YC2.1 with another circle of the same size, then pick the average signal values of the areas.

5. Move to Excel and subtract the background values from each of the YFP or CFP value in the corresponding channel to obtain crude signal values for YFP and CFP.

6. For each frame, divide YFP signal values with CFP signal values to obtain the ratio of YFP/CFP.

7. Calculate the baseline values (F_0 for YFP and CFP, R_0 for YFP/CFP, respectively) by averaging signal values of five frames before the stimulus onset, and subtract it from the raw signal values to obtain the signal changes (dF or dR). Calculate dF/F_0 or dR/R_0 (%), the values used for data quantification.

8. Data quantification can be done by taking the average of three frames around the peak value of dR/R_0, which is defined as the largest dR/R_0 value within 3 seconds after the stimulus onset (Fig. 2c).

4. Notes

4.1. Fly and Calcium Probes

As the baseline fluorescent intensity of the optical probe lowers, the signal to noise ratio becomes higher. The increased copy number of transgene constructs helps to enhance fluorescence. Accumulating the calcium indicator protein by rearing the fly longer (e.g., ~7 days) also helps to enhance fluorescence.

4.2. Recording Setup

- Fluorescent microscope: We use a standard upright fluorescent microscope (Axio Imager Z1, Carl Zeiss) equipped with electrical control of both the dichroic filter exchange and the shutter for excitation light. Motorizing these functions is not essential, but useful for avoiding possible disturbance of the devices surrounding the preparation in the dark room. It is preferable to customize the sample stage to create a larger space beneath the objective lens to facilitate the alignment of the flies and instruments. We also recommend to use a fixed stage system (e.g., Isolation System; Narishige).

- CCD camera and optics: Select a CCD camera with high sensitivity and a larger number of pixels with binning functionality that permits to increase the sensitivity. The dynamic range must be at least 12 bits.

- It is preferable to use dedicated computers for image acquisition via the CCD camera and infrared camera to ensure stable data acquisition.

- The least functionality required for the image processing is to pick signal values from region of interests (ROIs) in multi channel image stacks. Additional functionality such as image registration and spatial filtrations to reduce noises or movement artifacts would facilitate the analysis. We recommend ImageJ since it provides free and expandable environment for batch image processing.

4.3. Fly Preparation

- The legs of the male must be kept clean. Males with dirty legs are often found to be insistent on grooming and show weaker responses to sex-related stimuli. Bacterial breeding on the fly culture medium tends to be more significant in the vials with isolated adult males, whose legs have higher chances to carry sticky deposits. Changing the food vial every 1–2 days or adding antibiotics to the medium will facilitate troubleshooting of this problem. The flies that lost leg contacts with the styrofoam ball excrete more often waste matter, which could invite frequent glooming. It is therefore important to make the fly to hold the ball throughout the experiment.

5. Method

- Use a minimum amount of UV glue just sufficient to immobilize the fly. A thin tungsten wire would be useful for handling a small amount of glue. UV flash should be short (~5 s) to avoid damaging the fly with heat. Once the glue is applied, adjust the position of the fly as quickly as possible and cure the glue. The polyethylene wrap should be glued to the coverslip hole with tension, making it easier to cut a window on it.

- Minimize the time for cold anesthesia (~2 min).

- Keep watch if the fly in the humid chamber is holding the styrofoam ball. If you find a male without the ball, pick the ball with forceps and give it back to him to grab again.

- It is critical not to damage the air sacs. Tearing the air sacs often kill the fly in a few minutes.

- To avoid the photobleaching of the calcium sensor, minimize the lighting on the preparation once the brain is exposed.

- Avoid contaminations of cuticular hydrocarbons by cleaning the scissors and forceps with kimwipe wet with ethanol after the handling of each fly.

- Motorized control of leg-touching with microactuators might standardize the stimulus condition. However, we found that manual operation of the micromanipulator works best as it makes the contact stimulus gentle enough not to surprise the flies.

- Depending on the purpose and context of experiments, data analysis will use different parameters of a response, e.g., the peak amplitude, time course, and/or integrals.

Acknowledgments

This work was supported by MEXT grants 1802012 to D.Y. and 21770074 to M.K., a grant from the Strategic Japanese-French Cooperative Program funded by JST to D.Y., and a grant from the Tohoku Neuroscience GCOE program.

References

1. Brand AH, Perrimon N (1993) Targeted gene expression as a means of altering cell fates and generating dominant phenotypes. Development 118:401–415

2. Duffy JB (2002) GAL4 system in Drosophila: a fly geneticist's Swiss army knife. Genesis 34:1–15

3. Lee T, Luo L (1999) Mosaic analysis with a repressible cell marker for studies of gene function in neuronal morphogenesis. Neuron 22:451–461

4. Kitamoto T (2001) Conditional modification of behavior in Drosophila by targeted expression of a temperature-sensitive shibire allele in defined neurons. J Neurobiol 47:81–92

5. Hamada FN, Rosenzweig M, Kang K et al (2008) An internal thermal sensor controlling temperature preference in Drosophila. Nature 454:217–220

6. Sweeney ST, Broadie K, Keane J et al (1995) Targeted expression of tetanus toxin light chain in Drosophila specifically eliminates synaptic transmission and causes behavioral defects. Neuron 14:341–351

7. Marella S, Fischler W, Kong P et al (2006) Imaging taste responses in the fly brain reveals a functional map of taste category and behavior. Neuron 49:285–295

8. Gordon MD, Scott K (2009) Motor control in a Drosophila taste circuit. Neuron 61:373–384

9. Kitamoto T (2002) Conditional disruption of synaptic transmission induces male-male courtship behavior in Drosophila. Proc Natl Acad Sci U S A 99:13232–13237

10. Schroll C, Riemensperger T, Bucher D et al (2006) Light-induced activation of distinct modulatory neurons triggers appetitive or aversive learning in Drosophila larvae. Curr Biol 16:1741–1747

11. Suh GSB, Ben-Tabou de Leon S, Tanimoto H et al (2007) Light activation of an innate olfactory avoidance response in Drosophila. Curr Biol 17:905–908

12. Chiappe M, Seelig J, Reiser M et al (2010) Walking modulates speed sensitivity in Drosophila motion vision. Curr Biol 20(16): 1470–1475

13. Hendel T, Mank M, Schnell B et al (2008) Fluorescence changes of genetic calcium indicators and OGB-1 correlated with neural activity and calcium in vivo and in vitro. J Neurosci 28:7399–7411

14. Tian L, Hires SA, Mao T et al (2009) Imaging neural activity in worms, flies and mice with improved GCaMP calcium indicators. Nat Methods 6:875–881

15. Mank M, Santos AF, Direnberger S et al (2008) A genetically encoded calcium indicator for chronic in vivo two-photon imaging. Nat Methods 5:805–811

16. Nagai T, Yamada S, Tominaga T et al (2004) Expanded dynamic range of fluorescent indicators for Ca^{2+} by circularly permuted yellow fluorescent proteins. Proc Natl Acad Sci USA 101:10554–10559

17. Horikawa K, Yamada Y, Matsuda T et al (2010) Spontaneous network activity visualized by ultrasensitive Ca^{2+} indicators, yellow Cameleon-Nano. Nat Methods 7:729

18. Martin J-R, Rogers KL, Chagneau C et al (2007) In vivo bioluminescence imaging of Ca signalling in the brain of Drosophila. PLoS One 2:e275

19. Reiff D, Plett J, Mank M et al (2010) Visualizing retinotopic half-wave rectified input to the motion detection circuitry of Drosophila. Nat Neurosci 13:973–978

20. Buchner E (1976) Elementary movement detectors in an insect visual system. Biol Cybern 24(2):85–101

21. Borst A, Heisenberg M (1982) Osmotropotaxis in Drosophila melanogaster. J Comp Physiol 147:479–484

22. Kohatsu S, Koganezawa M, Yamamoto D (2011) Female contact activates male-specific interneurons that trigger stereotypic courtship behavior in Drosophila. Neuron 69:498–508

23. Antony C, Jallon JM (1982) The chemical basis for sex recognition in Drosophila melanogaster. J Insect Physiol 28:873–880

24. Diegelmann S, Fiala A, Leibold C et al (2002) Transgenic flies expressing the fluorescence calcium sensor Cameleon 2.1 under UAS control. Genesis 34:95–98

25. Fiala A, Spall T (2003) In vivo calcium imaging of brain activity in Drosophila by transgenic cameleon expression. Sci STKE 2003:PL6

26. Abramoff MD, Magelhaes PJ, Ram SJ (2004) Image processing with ImageJ. Biophoton Int 11:36–42

Chapter 7

Two-Photon Imaging of Population Activity with Genetically Encoded Calcium Indicators in Living Flies

Robert A.A. Campbell, Kyle S. Honegger, Eyal Gruntman, and Glenn C. Turner

Abstract

Genetically encoded calcium indicators make it possible to track neural activity on a population-wide level. Here we describe a preparation that enables two-photon imaging of neural activity in an essentially intact fly. We present strategies to minimize motion of the brain, both in preparation technique and in apparatus design. We discuss key variables for reducing the problems of photobleaching and phototoxicity in order to collect high quality imaging data. Finally, we discuss approaches to analyze the large quantities of data that can now be readily acquired using the latest generation of genetically encoded calcium indicators.

Key words: Calcium imaging, Two-photon microscopy, GCaMP, Mushroom body, Drosophila, Olfaction, Memory

1. Introduction

In the nervous system, it is typically the overall pattern of activity within a population of neurons that conveys information. For example, in the olfactory system, combinatorial patterns of activity are thought to convey information about the identity of an odor (1, 2). Population activity is also important in motor systems, where a particularly compelling example is the small network of neurons in the lobster stomatogastric ganglion which controls contractile movements of the gut (3). Each neuron fires a rhythmic pattern of action potentials; however, the pattern of no one neuron is responsible for the movements of the gut—individual neurons can fire identically when the gut muscles are moving in two different modes. Instead, it is the timing of each neuron's firing relative to the others that dictates the gut movements, a feature that is only visible when examining activity of the population as a whole.

Jean-René Martin (ed.), *Genetically Encoded Functional Indicators*, Neuromethods, vol. 72,
DOI 10.1007/978-1-62703-014-4_7, © Springer Science+Business Media, LLC 2012

Optical methods hold great promise for monitoring population-level activity. Recent developments in genetically encoded calcium indicators (GECIs (4–7)) are sure to accelerate our understanding of neural circuit function in Drosophila. The genetic toolkit of Drosophila enables fly biologists to target these indicators to particular cell types, subsets of cell types and even individual cells, with a level of precision and reproducibility unrivaled by other model systems (8–10). The sensors report changes in calcium concentration that accompany neural activity either as increases in fluorescence or as changes in fluorescence resonance energy transfer (FRET) (11). Since neural activity can drive changes in calcium concentration over a wide range (12), GECIs have the potential to display the high signal-to-noise ratios needed to visualize small changes in neural activity. The early generations of GECIs had fairly low sensitivity, and only detected calcium concentrations at the upper range of fluctuations that occur in vivo. However, a new sensor, GCaMP 3.0, a circularly permuted GFP with a calmodulin-controlled fluorophore environment, has dramatically improved responsiveness, approaching single-spike sensitivity in some experimental settings (5). Unfortunately, GECIs still suffer from slow temporal dynamics, with rise times of about 100 ms and decay times of over 600 ms for GCaMP 3.0 (5).

Despite the limitations of sensitivity and temporal resolution, GECIs are extremely useful for monitoring neural activity on a large scale. For example, they can provide an overview of activity in a population of neurons identified using a GAL4-UAS-shibire based dissection of behavior. This overall view can be refined with electrophysiological studies to investigate neuronal activity at higher resolution. Our particular interest has been in characterizing sensory responses in the mushroom body (MB), an olfactory learning and memory center of the fly brain (13, 14). Electrophysiological recordings have shown that the intrinsic neurons of the mushroom body (Kenyon cells or KCs) have highly odor-selective responses (15, 16). This selectivity is thought to be important for the accuracy of memory formation (17, 18). The MB is a particularly good venue for applying imaging techniques to study population coding because individual neurons have extremely low baseline spiking rates, but when they do respond, they fire several spikes. This type of activity is well captured by GCaMP 3.0, since it has a high sensitivity.

There are a few important technical issues to consider when monitoring neural activity with imaging. The first is photobleaching and phototoxicity. While these are both an inevitable aspect of imaging, phototoxicity is drastically reduced when using two-photon imaging methods rather than single photon techniques, i.e., laser-scanning confocal microscopy or wide-field imaging with an incandescent lamp and CCD camera (19, 20). The wavelengths of light used for two-photon imaging (920 nm for GFP-based imaging) are absorbed much less by the endogenous chromophores in neural tissue, so the photons both penetrate tissue more deeply and cause less photodamage than wavelengths necessary for single photon excitation (21). However, although fluorophore excitation is confined to a small region of the tissue due to the two-photon effect, one should bear in mind that there are still many single photons in the infrared range that are heating the preparation. Nevertheless, for in vivo preparations, two-photon excitation is vastly superior to single-photon methods, where more damaging wavelengths are applied to the entire depth of the sample only to recover fluorescence emission from a small region of the tissue.

A second consideration is that GECIs report changes in calcium concentration, not the spikes that are the actual currency of communication in the nervous system. Calcium influx is coarsely correlated with spiking, but calcium can enter the cell either through voltage-gated calcium channels, or through ligand-gated (i.e., synaptic) ion channels (22). This raises the possibility that synaptic input could potentially be mistaken for spiking output. This possibility is a particular concern in insects where most central synaptic transmission is through nicotinic acetylcholine receptors, which are significantly calcium permeable (23). The different sources of calcium influx likely account for results where strong signals are observed in dendrites, with only weak activity in the somata (24, 25). The relationship between fluorescence changes and spiking activity can be established by simultaneous electrophysiological recordings and imaging. Such calibrations are typically done using synaptic stimulation to drive various levels of spiking activity in the neurons of interest and so generate a calibration curve (4, 26, 27).

The relationship between the spiking activity of a cell and the fluorescence changes one measures will be specific for each cell type. This is because the relationship depends on: (1) the concentration of GCaMP inside the cell; (2) the relative contribution of ligand-gated versus voltage-gated calcium channels to activity dependent changes in calcium influx. For example, one type of neuron may express very low levels of a voltage-gated calcium channel, so most of the calcium influx to the neuron reflects synaptic input rather than spiking output, while the reverse could be true in another neuron. Another factor is which cellular compartment one images. Calcium changes in the cell body are thought to derive from activation of voltage-gated channels, and therefore represent spiking output, while imaging from dendritic sites (which may be prominent structures with useful morphological features) may more heavily reflect synaptic input. Moreover, since the level of GCaMP expression is also an important factor, there could be a different relationship between spikes and reporter signal even in the same cell type depending on the strength of the GAL4 driver. There are no broadly applicable rules; however, these factors should be considered when interpreting signals generated by calcium sensors.

Calibrating the relationship between spikes and fluorescence signals can enable one to reconstruct the rate and time course of action potential firing from the acquired florescence signal (27). The key aspect here is that one must establish the time course of calcium signal in response to a single spike. The fluorescence changes to a spike train are essentially a convolution of this impulse response with the spike train. Consequently, knowing the impulse response to a single spike, one can deconvolve the fluorescence time course into a spiking rate time course (27).

The third challenge is the purely technical difficulty of minimizing the motion of the tissue of interest. Most experiments require tracking the same brain region or the same cells over time. This can be difficult within a living organism where muscular movements can cause large motion artifacts, and even slow drift can cause one to lose track of a particular region of interest over time. Here we describe a preparation that enables imaging from neurons of interest while leaving the fly essentially intact (see also (28)). We also provide guidelines and discuss key variables for acquiring high quality data, based on our experience of imaging Kenyon cells. However, each investigator will have slightly different requirements, and will have to adapt these methods for their own purposes.

In Sect. 3.1 below, we discuss methods of constructing a recording platform to hold the fly while minimizing problems from vibration and slow drifting movements. Section 3.2 details the dissection procedure for exposing neurons of interest while still being able to deliver sensory stimuli in a living fly. In Sect. 3.3 we describe the relevant variables for obtaining good quality results, both generally and with an emphasis on recording from KCs. And Sect. 3.4 gives an introduction to basic analyses for interpreting the data. GECIs provide an exciting new way of probing neural function in genetic model organisms. Although there are some limitations, researchers can now target expression of sensors to neural populations implicated in particular behaviors, and get a meaningful and high-resolution view of neural codes in action (29).

2. Materials

2.1. Reagents

Bath saline
NaCl 103 mM
KCl 3 mM
$CaCl_2$ 1.5 mM
$MgCl_2$ 4 mM
NaH_2PO_4 1 mM
$NaHCO_3$ 26 mM

N-Tris(hydroxymethyl)methyl-2-aminoethanesulfonic acid 5 mM

Trehalose 10 mM

Glucose 10 mM

1. The osmolarity of the solution should be 275–280 mOsm. If necessary adjust by adding sucrose; sucrose is not metabolized by the brain and acts only as an inert solute.

2. Store at 4°C to minimize bacterial growth.

3. Before use bring to room temperature and oxygenate by bubbling 95% O_2/5% CO_2 continuously through the solution. The pH should equilibrate to 7.3.

2.2. Equipment

Sharpening Stone

Forceps (Roboz by Dumont, #5 INOX, BIOLOGIE tip), further sharpened by hand for dissection.

Forceps (Roboz by Dumont #3), blunt to manipulate fly into recording platform without damage.

Epoxy (Devcon 5-min epoxy)

Dissecting microscope with ~150× total magnification

Two-photon imaging system

95% O_2–5% CO_2 gas tanks with regulator

---

3. Methods

3.1. Constructing a Recording Platform

The fly will be inserted into a hole cut into a piece of aluminum foil. This immobilizes the head of the fly, and permits the exposed neurons to be bathed in saline above the surface of the foil, while the sensory structures, including the antennae, proboscis, maxillary palps and much of the eyes, remain dry below the surface of the foil. The foil is located in a well on a larger platform that allows bath saline to be perfused over the fly, and enables one to easily mount the preparation on the microscope.

Platforms will have to be adapted to each user's imaging apparatus; the platform we use is shown in Fig. 1a. Important considerations for the design are the following:

1. To minimize movement, the platform should be made of relatively rigid material and held tightly in place on the microscope stage. We use a disk-shaped platform made out of hard plastic that affixes to the microscope stage with small pegs.

2. Any stimulus delivery apparatus should not be connected to the platform directly, but to a more stable point on the stage or airtable.

3. The well where the fly sits should be relatively shallow (~2 mm deep) to ensure complete exchange of perfusion over the

a   Recording Platform

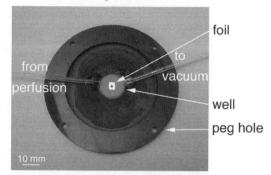

b Mounting the Fly and Dissection

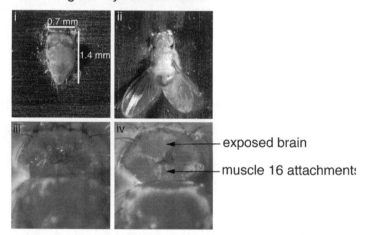

Fig. 1. (**a**) Recording platform. Constructed of rigid plastic, with holes around the outer edge for positioning platform on pegs on microscope stage. To form the well, a circular section is removed from the middle of the platform, and a plastic weigh boat is epoxied to the underside. A small square hole is cut in the weigh boat, and the foil glued underneath. The hole for the fly is then fashioned using a fine gauge syringe needle. (**b**) (i) Fly inserted into the foil holder in the recording platform, viewed from the top of platform (dorsal side of fly). (ii) View from the underside of platform (ventral side of fly). The fly's legs have been removed in order to take the photo. (iii) Dorsal side of the fly after being fixed in place. The head is covered in bath saline. (iv) The cuticle has been opened to reveal the brain, the fat and air sacs/trachea have been moved aside. Notice that only a section of cuticle has been removed to reveal the brain region containing the cells of interest, in this case, mushroom body Kenyon cells. The location of muscle 16 attachments is indicated. Panel (**b**) © CSHL Press.

preparation. If the well is too deep, saline can pool and the dissolved oxygen will form bubbles under the objective lens.

The hole in the foil can be formed using a 25 gauge needle; shape and dimensions are indicated in Fig. 1a (see Note 1). Small adjustments to the shape of the opening may be required depending on the size of the flies used for recording (e.g., males or females). The foil will deform after several preparations, and will need to be frequently replaced.

3.2. Dissection

1. Transfer a small number of flies to a glass sample tube and anesthetize on ice until movement ceases (about 15 s).

2. Using blunt forceps carefully transfer a fly to the recording platform. We usually use females since they are larger than males and so easier to dissect. Picking flies up by a wing minimizes the possibility of damage.

3. Insert the fly through the underside of the foil hole cut in the platform. Seat the fly into the hole by pressing against the thorax with one forceps while wedging the sides of the thorax into the hole with the other. Abort the dissection if damage to the cuticle causes fluid to leak out, in which case the fly will soon dehydrate and die. Once the thorax is inside the hole, flip the platform and gently lift the head through the hole by pushing the thorax backwards. If imaging KC cell bodies, tilt the fly's head forward to provide access to the posterior surface of the brain where the KC somata are located. The olfactory organs point downwards in this preparation, allowing airborne odor delivery from the underside of the platform.

4. Once a good position is achieved, fix the fly in place using fast-drying epoxy (Devcon 5-min epoxy). The head must be firmly fixed in place for subsequent dissection steps. Since it is not strongly adhesive, the epoxy should be shaped over and around the cuticle so that when it hardens it will prevent the head capsule from pulling loose as pieces of cuticle are plucked off. Applying a small quantity of epoxy between the head and thorax (i.e., to the sides of the neck) can effectively stabilize the head (see Note 2). It is not strictly necessary to epoxy the thorax to the foil. The rest of the fly can remain simply mechanically wedged in place; leakage of saline through the gaps between the platform and the fly's abdomen are a rare occurrence since surface tension generally prevents the saline from passing through these small areas.

5. Pumping movements of the brain can prevent reliable data collection. These movements can be minimized by removing muscle 16, which connects the aortic funnel to the frontal pulsatile organ (30). This muscle can be removed from the front of the head (dry side of the platform) by tearing a small hole in the cuticle between the two antennae at the level of their base, inserting the forceps to extract the muscle, which will appear as a silvery thread. This part of the dissection depends critically on well-sharpened #5 forceps whose points meet well.

6. Fill the platform well with bath saline and dissect away the cuticle at the back of the head, being careful not to pull the head free of the epoxy. Gentle peeling motions work well. Air sacs and fat deposits lying over top of the area of interest should be gently moved to the side, but removing air sacs fully can allow the brain to move more within the head capsule.

We often remove the proboscis retractor muscles, which pass over the caudal aspect of the optic lobes. If you had trouble removing muscle 16 via the front of the head, it may be possible to extract it in this orientation. It can be removed at the neck, just dorsal to the gut; this is easiest if the head is tilted forwards, stretching out the neck. We do not remove the perineural sheath for imaging. Periodically exchange the bath with fresh oxygenated saline at 2–3 min intervals throughout the dissection.

7. Once the dissection is complete, transfer the fly to the microscope stage. Place the perfusion and vacuum lines so the animal can be continuously perfused with oxygenated saline at a flow rate of ~2 mL/min for the duration of the experiment. Take care that the vacuum line does not become plugged and is properly positioned to prevent saline leaks.

3.3. Imaging

We use a Prairie Ultima system (Prairie Technologies) that comes with a motorized stage we have customized to hold our recording platform. Pegs fitted into the motorized stage insert into holes in the recording platform to increase stability and ensure consistent positioning of the preparation. The stage and heavy airtable, together with the platform peg-anchors, minimize movement from both vibration and slow drift.

Obtaining high quality results with two-photon imaging is largely a matter of minimizing photobleaching and phototoxicity, while maintaining good signal-to-noise ratio throughout the duration of the imaging session. Selecting a power level and the overall imaging strategy will be specific for each application (see Note 3). Here we offer guidelines that are useful for imaging KCs. These procedures are designed to allow long imaging sessions to present multiple stimuli in an experimental context with a strong expression of GCaMP 3.0 in small cells.

To track individual neurons we image from the cell body layer in frame-scan mode using the lowest power settings that produce an acceptable signal-to-noise ratio throughout the recording session. Optimal power settings will be different for individual applications; relevant factors are the level of GECI expression, depth of imaging plane, and required duration of the imaging experiment. Phototoxicity increases substantially at laser intensities of over 15 mW at the back aperture of the objective (31), but with a strong GAL4 driver, good results can be obtained with power settings no more than 8–10 mW.

The signal-to-noise ratio can be enhanced by averaging, either by increasing the dwell time of the laser on individual pixels or by averaging across multiple frames. For imaging sessions longer than 15 min we find that a useful strategy is to use short pixel dwell times and instead average across multiple frames. This strategy helps to compensate for brain movements and deformations over time, since the amount of cellular movement during the acquisition of individual frames is relatively small when the frame rate is high. With this overall approach, it is reasonable to expect imaging sessions lasting 20 min where individual cells can be tracked throughout and signal-to-noise ratio remains at a high level. Preparations with larger cells, located deeper in the brain, can last substantially longer than 20 minutes. The limiting factor is usually cell movements, particularly in the z-axis, which prevent us from tracking cells.

3.4. Analysis

In most experiments, stimuli are presented individually in "trials" during which fluorescence time-series data are acquired. These trials must consist of a baseline period of several frames when no stimulus is given and several frames of response period during and after stimulus presentation. In vivo imaging is often associated with substantial brain motion and the region of interest can easily drift out of the field of view. It may, therefore, be necessary for line-scans to either be supplemented with or replaced by frame-scans. Whilst frame-scans are slower than line-scans, they allow for the possibility of motion-correction in many cases. Motion-correction can either be achieved by manually adjusting x, y, or z position online or, in some cases, by translation-correction algorithms offline (32).

Imaging software generally stores data as single or multi-frame TIFF (tagged image file format) images. Large volumes of data are created and these may require extensive processing. Automated analysis techniques are a must. Matlab is commonly used since it is easy to learn, well documented, and has a large suite of image processing functions. Cost-free and open-source alternatives include Octave, Python and its associated analysis environment, Sage, and Perl with its associated environment, PDL. Finally, R is useful for fitting statistical models.

The analysis steps for imaging data are straightforward. If frame-scan data were obtained then the first step is to correct motion artifacts by ensuring that all frames are aligned. Assuming the brain moves little during the course of a single frame then a simple translation-based correction along the x–y domain is usually sufficient (32). If a large degree of motion occurs in z, perpendicular to the imaging plane, then it may not be possible to register images and the data must be discarded. Motion artifacts will be less severe at higher frame rates since the brain will not have moved significantly during acquisition of a single frame. The degree to which motion can be tolerated will depend upon the pattern of GCaMP expression and the goals of the experiment.

The proportional evoked change in fluorescence (the dF/F) is calculated with respect to a period of baseline activity. In order to get the most accurate measure possible it is desirable to subtract the offset due to baseline signal: that which does not arise from the indicator. Shot noise, tissue autofluorescence, and scattered fluorescence could all contribute to this baseline. To estimate the contribution of shot noise alone, one can acquire a small number of frames at the start of each trial with the laser shutter closed. Alternatively, one can subtract both shot noise and autofluorescence by subtracting the mean fluorescence in regions of the image where GCaMP is not expressed. In this case, there is no need to acquire frames with the shutter closed.

At this stage, it is now possible to select a region of interest within the image (cells bodies or regions of neuropil) and calculate the associated mean change in fluorescence (dF/F) over time. The resulting dF/F time-course can be summarized further by integrating the area under the response peak. In this manner it is potentially possible to record the activity of hundreds of neurons simultaneously. The response of each neuron on each trial can be reduced to a single mean evoked dF/F value.

4. Notes

1. The size of the hole is critical for ensuring good positioning of the fly and an easy dissection. We find that a hole with a width of 0.7 mm and a length of 1.1 mm is optimal for female flies. However, culturing conditions and genotype can influence fly size significantly at this scale. The hole must be tight enough to hold the fly's head fixed as it is repositioned, but not so tight as to distort the head.

2. When applying the epoxy, deposit it on the foil between the head and the thorax and then gently smear it over the eye. Do not deliver the drop of epoxy straight onto the fly since this will displace the fly from the desired position. It is best to wait until the epoxy has begun to become more viscous before applying. This minimizes the risk of epoxy wicking underneath the platform and coating the fly's underside and legs.

3. Scanning considerations: For two-photon and other laser scanning approaches, one moves the excitation source over a series of coordinates that become the pixels of the image. How the experimenter chooses to collect those pixels represents a balance between high temporal resolution and signal detection in the presence of noise. Spending less time exciting a particular pixel volume (referred to as the pixel dwell time) will increase the number of pixels one may acquire in a certain amount of time, and thus the temporal resolution of the data. However, this is offset by the fact that most sensors do not produce an extremely large amplitude signal—dF/F values are only 2 or 3 at maximum, so on short timescales noise

overwhelms the signal. Since most noise in a two-photon system is stochastic over time (shot noise), averaging over a sufficiently long time window will effectively separate signal from noise. To achieve this, one can either average a pixel value within a scan using longer pixel dwell times, or average the same pixel across multiple scans. In some cases, the second strategy is preferable, imaging at high frame rates using the minimal dwell time per pixel and then averaging the signal across more scans. This yields a noisier signal per frame, but it is possible to average over frames. More importantly, high frame rates can be helpful when tracking individual neurons, where minimizing the duration of a frame enables one to more easily track slowly moving cells across successive scans. However, different experimental scenarios will call for different strategies for spreading out pixel collection time (line-scan vs. frame-scan; short dwell, more scans vs. long-dwell, fewer scans; etc.) in a way that optimizes both temporal resolution and signal-to-noise.

Acknowledgments

K.S.H. is supported by the Crick-Clay fellowship from the Watson School of Biological Sciences and predoctoral training grant 5T32GM065094 from the National Institute of General Medical Sciences. E.G. is supported by the Elisabeth Sloan Livingston fellowship from the Watson School of Biological Sciences. This work was funded by NIH grant R01 DC010403-01A1.

References

1. Malnic B, Hirono J, Sato T, Buck LB (1999) Combinatorial receptor codes for odors. Cell 96:713–723

2. Su C-Y, Menuz K, Carlson JR (2009) Olfactory perception: receptors, cells, and circuits. Cell 139:45–59

3. Marder E, Bucher D (2007) Understanding circuit dynamics using the stomatogastric nervous system of lobsters and crabs. Annu Rev Physiol 69:291–316

4. Reiff DF, Ihring A, Guerrero G, Isacoff EY, Joesch M, Nakai J, Borst A (2005) In vivo performance of genetically encoded indicators of neural activity in flies. J Neurosci 25: 4766–4778

5. Tian L, Hires SA, Mao T, Huber D, Chiappe ME, Chalasani SH, Petreanu L, Akerboom J, McKinney SA, Schreiter ER, Bargmann CI, Jayaraman V, Svoboda K, Looger LL (2009) Imaging neural activity in worms, flies and mice with improved GCaMP calcium indicators. Nat Methods 6:875–881

6. Miyawaki A, Llopis J, Heim R, McCaffery JM, Adams JA, Ikura M, Tsien RY (1997) Fluorescent indicators for Ca^{2+} based on green fluorescent proteins and calmodulin. Nature 388:882–887

7. Mank M, Santos AF, Direnberger S, Mrsic-Flogel TD, Hofer SB, Stein V, Hendel T, Reiff DF, Levelt C, Borst A, Bonhoeffer T, Hübener M, Griesbeck O (2008) A genetically encoded calcium indicator for chronic in vivo two-photon imaging. Nat Methods 5:805–811

8. Brand AH, Perrion N (1993) Targeted gene expression as a means of altering cell fates and generating dominant phenotypes. Development 118:401–415

9. Lai S-L, Lee T (2006) Genetic mosaic with dual binary transcriptional systems in Drosophila. Nat Neurosci 9:703–709

10. Potter CJ, Tasic B, Russler EV, Liang L, Luo L (2010) The Q system: a repressible binary system for transgene expression, lineage tracing, and mosaic analysis. Cell 141:536–548

11. Mank M, Griesbeck O (2008) Genetically encoded calcium indicators. Chem Rev 108:1550–1564

12. Stosiek C, Garaschuk O, Holthoff K, Konnerth A (2003) In vivo two-photon calcium imaging of neuronal networks. Proc Natl Acad Sci USA 100:7319–7324

13. Davis RL (2005) Olfactory memory formation in Drosophila: from molecular to systems neuroscience. Annu Rev Neurosci 28: 275–302

14. Keene A, Waddell S (2007) Drosophila olfactory memory: single genes to complex neural circuits. Nat Rev Neurosci 8:341–354

15. Perez-Orive J, Mazor O, Turner GC, Cassenaer S, Wilson RI, Laurent G (2002) Oscillations and sparsening of odor representations in the mushroom body. Science 297:359–365

16. Turner GC, Bazhenov M, Laurent G (2008) Olfactory representations by Drosophila mushroom body neurons. J Neurophysiol 99:734–746

17. Kanerva P (1988) Sparse distributed memory. MIT Press, Cambridge, MA

18. Marr D (1969) A theory of cerebellar cortex. J Physiol (Lond) 202:437–470

19. Yuste R (2005) Imaging in neuroscience and development: a laboratory manual. Cold Spring Harbor Laboratory Press, Cold Spring Harbor, NY

20. Grewe BF, Helmchen F (2009) Optical probing of neuronal ensemble activity. Curr Opin Neurobiol 19:520–529

21. Helmchen F, Denk W (2005) Deep tissue two-photon microscopy. Nat Methods 2: 932–940

22. Oertner TG, Single S, Borst A (1999) Separation of voltage- and ligand-gated calcium influx in locust neurons by optical imaging. Neurosci Lett 274:95–98

23. Single S, Borst A (2002) Different mechanisms of calcium entry within different dendritic compartments. J Neurophysiol 87: 1616–1624

24. Wang Y, Wright NJ, Guo H, Xie Z, Svoboda K, Malinow R, Smith DP, Zhong Y (2001) Genetic manipulation of the odor-evoked distributed neural activity in the Drosophila mushroom body. Neuron 29:267–276

25. Wang Y, Guo H-F, Pologruto TA, Hannan F, Hakker I, Svoboda K, Zhong Y (2004) Stereotyped odor-evoked activity in the mushroom body of Drosophila revealed by green fluorescent protein-based Ca^{2+} imaging. J Neurosci 24:6507–6514

26. Jayaraman V, Laurent G (2007) Evaluating a genetically encoded optical sensor of neural activity using electrophysiology in intact adult fruit flies. Front Neural Circ 1:3

27. Yaksi E, Friedrich RW (2006) Reconstruction of firing rate changes across neuronal populations by temporally deconvolved Ca^{2+} imaging. Nat Methods 3:377–383

28. Murthy M, Turner GC (2010) In vivo whole-cell recordings in the Drosophila brain. In: Zhang B, Waddell S, Freeman M (eds) Drosophila neurobiology methods: a laboratory manual. Cold Spring Harbor Laboratory Press, Cold Spring Harbor, NY

29. Wang JW, Wong AM, Flores J, Vosshall LB, Axel R (2003) Two-photon calcium imaging reveals an odor-evoked map of activity in the fly brain. Cell 112:271–282

30. Bate M (1993) The development of Drosophila melanogaster. Cold Spring Harbor Laboratory Press, Cold Spring Harbor, NY

31. Seelig JD, Chiappe ME, Lott GK, Dutta A, Osborne JE, Reiser MB, Jayaraman V (2010) Two-photon calcium imaging from head-fixed Drosophila during optomotor walking behavior. Nat Methods 7:535–540

32. Guizar-Sicairos M, Thurman ST, Fienup JR (2008) Efficient subpixel image registration algorithms. Opt Lett 33:156–158

Chapter 8

Engineering and Application of Genetically Encoded Calcium Indicators

Jasper Akerboom, Lin Tian, Jonathan S. Marvin, and Loren L. Looger

Abstract

Genetically encoded fluorescent biosensors are useful tools for tracking target analytes in cells, tissues and living organisms. These probes are often chimeric proteins consisting of a recognition element (e.g., a ligand-binding protein) and a reporter element (one or more fluorescent proteins). The analyte-induced conformational change in the recognition element leads to an observable change in fluorescence in the reporter element. Expression of biosensors is noninvasive and can be targeted to specific tissues and cell types using specific promoter and enhancer sequences, and to subcellular compartments with signal peptides and retention tags. Recent improvements in both indicator engineering and microscopy methods enable chronic *in vivo* measurements. Here, we describe methods used in the design, testing, optimization and application of genetically encoded biosensors, with a particular focus on the widely utilized calcium indicator GCaMP.

Key words: GCaMP, GECI, Calcium imaging, Neural activity imaging, GCaMP3, Protein engineering

1. Introduction

In general, a protein-based biosensor contains a recognition element and a reporter element. Recognition elements of biosensors frequently take advantage of naturally evolved or selected proteins or peptides that specifically bind to a target analyte with affinities in the range of physiological concentrations. Reporter elements, such as a covalently attached dye or a fused fluorescent protein, transduce the analyte-induced conformational change in the recognition element into an observable signal, such as an increase in fluorescence emission. Protein engineering is the adaptation and selection of polypeptides for a desired functionality (1, 2) and has been used widely in the development of protein-based sensors.

Jean-René Martin (ed.), *Genetically Encoded Functional Indicators*, Neuromethods, vol. 72,
DOI 10.1007/978-1-62703-014-4_8, © Springer Science+Business Media, LLC 2012

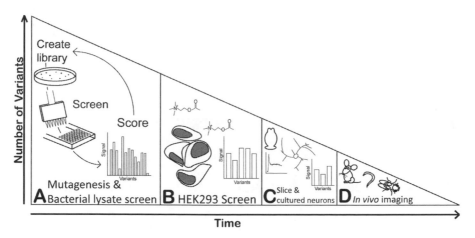

Fig. 1. Schematic overview of the GECI screen. (**a**) Sensor mutagenesis and bacterial lysate screen: this step can generate the largest diversity, and is often the first step in GECI optimization. Many variants can be tested, but this screen will generate many false positives. (**b**) HEK293 screen: promising variants are fed into this more stringent screen, and false positives and slow responders can be discarded. (**c**) Cultured neurons and slice: GECI candidates passing the HEK293 screen are tested for longer-term expression effects and response to spontaneous or evoked action potentials. (**d**) The best variant(s) from the slice and cultured neuron assays are finally subjected to *in vivo* imaging in model organisms (either spontaneous, evoked, or behaviorally correlated activity).

Below we describe in detail the principles of sensor construction, optimization and screening (see Fig. 1). We also discuss considerations for downstream sensor use, including *in vivo*.

2. Protein Engineering Techniques

2.1. Rational Design and Directed Evolution

Several protein-engineering techniques are described in the literature, and they can be roughly grouped into two categories: rational design (3) and directed evolution (4). Rational design has evolved over the years from "design by inspection" (5) and sequence grafting (6) to rely more on sophisticated models of atomic interactions (7), combinatorial optimization algorithms (8) and advanced computational power (9). Recent examples of successful designs include de novo design of a protein fold (10) catalytic activity (11, 12) and alteration of ligand binding (11–14). Substantially redesigned proteins often possess low levels of activity and poor thermodynamic stability (11, 13), are sometimes prone to aggregation, and can function unexpectedly in settings other than those in which they were screened (15). In these cases, directed evolution may improve such properties (16), illustrating that there remain many terms not adequately captured by modeling methods. Explicitly combined design and evolution is emerging as the most powerful method, with simple protocols (17). However, in simple cases the success rate can be high, especially with the advent of modern "design by inspection" viewers such as PyMol (http://www.pymol.org), Swiss PDB Viewer (http://spdbv.vital-it.ch), Yasara (http://www.yasara.org), and KiNG (http://kinemage.biochem.duke.edu) (for a review of molecular visualization see ref. (18)).

Directed evolution couples generation of protein sequence diversity with a screen (19) (every variant is analyzed, e.g., fluorescence) or a selection (20) (only variants surpassing a survival criterion are analyzed, e.g., antibiotic resistance). Directed evolution can be easy to implement, even in the absence of meaningful structural information (4). Typically, iterative rounds of stringency, sequencing, analysis, and library diversity generation are employed to reach final variants optimizing design criteria (21). Numerous methods for generating DNA diversity are available (22–26), each with strengths and weaknesses, summarized in Table 1. In general, directed evolution can be quite successful, but potential difficulties exist, such as the inability to select for a desired trait (e.g., a fluorescent biosensor), large changes in sequence space required to produce the desired activity (e.g., dramatic substrate change for enzymes), incompatibility with screening system (e.g., posttranslational modifications and codon bias), and insoluble or toxic products (e.g., membrane proteins, accumulation of metabolic intermediates). Also, bias during library generation and difficulties during candidate screening and selection may skew the results (17).

2.2. Datamining, Model Building, Sequence Analysis

Detailed knowledge about the protein one wants to engineer is valuable, therefore doing extensive data mining in literature, sequence and structure databases is an essential part of any project. Protein-specific databases of note are Brenda (http://www.brenda-enzymes.org) (27), CAZy (http://www.cazy.org) (28), Swiss-Prot (http://expasy.org/sprot) (29), iHOP (http://www.ihop-net.org) (30), and the wwPDB (http://www.wwpdb.org) (31); many more tools and databases are available online. If structural information on the scaffold protein is not available, one can compute a three-dimensional model using Web servers available for remote homology/fold recognition, such as Phyre (http://www.sbg.bio.ic.ac.uk/~phyre) (32). If a structural homologue is absent and structure determination is not an option, a 3D model can be built de novo using, for example, the I-Tasser (http://zhanglab.ccmb.med.umich.edu/I-TASSER) (33) or Robetta (http://robetta.bakerlab.org) (34) Web servers. Direct sequence comparison methods have been successfully applied to guide protein-engineering strategies as well (35, 36).

2.3. Choice of Method

After sufficient information has been gathered, one must next choose the method(s) to reach the target objective. Small changes, such as an increase in optimum temperature for a certain enzymatic/binding activity, or a minor alteration in substrate specificity or fluorescence, might be achieved by performing a few rounds of

Table 1
**Directed evolution and screening methods available for protein engineering
(list is not meant to be complete)**

Technique	Description	Advantages	Disadvantages
Error prone PCR (116)	Suboptimal PCR, DNA mismatches are created by DNA polymerase	No structural information is needed, easy, quick and cheap to perform	Bias for certain mutations (e.g., A to T, C to G) Library size can be prohibitively large, with many loss-of-function mutations
Kunkel mutagenesis (43)	Plasmid template containing uridine instead of thymine is copied using a primer containing a specific mutation	In theory many mutations can be generated in one round, quickly and cheaply. Deletions and insertions are also possible	Plasmid needs to contain an f1 ori, and multiple steps and bacterial strains are needed to generate the uracil template
Ribosome and mRNA display (117, 118)	*In vitro* selection of protein libraries using proteins bound to either mRNA or ribosomes and mRNA for specific binding targets	Library diversity can be large due to *in vitro* selection, and can be combined with PCR mutagenesis	Procedure is only suited for soluble proteins and screening is only possible for binding, not for catalysis. Not clear if sensors will function properly when displayed
Phage display (46, 119–121)	Proteins and peptides are displayed on filamentous phage and can be extracted from large populations *in vitro*	Libraries can be very large, phage genome is small and easy to handle, candidates are easily selected and sequenced	Phage display is only suited for binding not for catalysis, proteins should be fairly small. Many proteins do not display well; not clear if sensors will function properly when displayed
DNA shuffling (23, 40)	Genes encoding different proteins are reassembled in a PCR-like reaction, resulting in chimeras of the different templates	Compared to other directed evolution methods, bigger changes in sequence space can be generated	The different genes used in the assembly reaction should share a sequence identity ~70% for the mutagenesis to be successful

directed evolution, potentially guided by structural information (37, 38). When greater changes are required (complete alteration of substrate specificity, generating new activity from scratch), rational design methods followed by directed evolution have the greatest chance of success (17). Important to note is that for directed evolution, the screen/selection employed predetermines the best possible outcome (i.e., "you get what you screen for" (39)). Therefore, if screening is difficult or impossible to perform adequately, initial rational design methods should be applied.

2.4. Targeted Mutagenesis

As described above, many different methods for generating libraries for analysis are available. DNA shuffling (40) and other techniques (24, 26) have produced extraordinary results (23, 37, 41, 42), but are more amenable to selections of enzymatic activity due to the high fraction of nonfunctional sequences produced by widespread recombination. We require a method of site-directed mutagenesis that is efficient for substitutions, insertions, and deletions, and that can be applied to multiple positions simultaneously. Of the techniques listed, we find single-stranded uracil template ("Kunkel") mutagenesis (43) to best meet our criteria. We have optimized the protocol, which we present briefly here and in Sect. 2.4.1 (Protocol 1).

During Kunkel mutagenesis, one replicates a template vector containing the gene to be mutagenized with primers containing specific mutations, insertions or deletions. The technique can be used for library generation with degenerate primers, and is compatible with multi-site mutagenesis, although efficiency declines with increasing complexity (44). The template vector contains uridine instead of thymine at random positions in the vector, resulting in degradation of the template DNA when transformed to *E. coli* cloning strains such as XL-1 and DH5α.

2.4.1. Protocol 1: Kunkel Mutagenesis

A plasmid template for Kunkel mutagenesis must contain, in addition to an origin of replication (ori) for double stranded DNA replication, a bacteriophage ori (e.g., f1). Such a plasmid, termed a phagemid, is transformed into a special *E. coli* strain, CJ236, deficient in two enzymes, UTPase and uracil-DNA-glycosidase, and containing a helper plasmid pCJ105, encoding the F-pilus (45). The plates used during transformation must be selective for both the antibiotic resistance encoded on the phagemid as well as the antibiotic present on the plasmid pCJ105 (i.e., chloramphenicol). The following day, a colony is selected and grown at 37°C, shaking, in 3 ml LB medium containing the antibiotic appropriate to the phagemid and chloramphenicol. During late log phase, when the culture is visibly turbid, the helper phage M13K07 is added to approximately 2×10^8 plaque forming units (pfu)/ml, and the culture is allowed to incubate for another hour at 37°C, shaking, after which 100 ml LB is added, supplemented only with the appropriate antibiotic for phagemid selection. The culture is then incubated overnight at 37°C, shaking. Addition of kanamycin to a final concentration of 100 µg/ml to select for M13K07 infected cells during this step tends to increase phage yield, but sometimes at the expense of quality of the single-stranded uratidyl-DNA isolated. Dividing the culture in two, and adding kanamycin to only half of the culture is a good strategy.

The following day, phage particles are isolated by repeated PEG/NaCl precipitation, performed as follows. Bacteria are pelleted by centrifugation, at $8,000 \times g$ for 10 min at 4°C. The supernatant is transferred to sterile conical tube(s), and 20% v/v ice-cold

Phage Precipitation Buffer (20% w/v PEG 8000, 2.5 M NaCl) is added. After thorough mixing, this solution is allowed to incubate on ice for 10 min; the precipitating phage should make the supernatant slightly cloudy. The phage is then pelleted at $8,000 \times g$ for 10 min at 4°C. The supernatant is removed from the faintly visible pellet by careful decanting and pipetting. The phage pellet is subsequently resuspended in 2 ml PBS buffer by vortexing, and transferred to two reaction tubes. A 5-min spin at maximum speed in a tabletop centrifuge will pellet bacteria that might have been carried over (double-stranded bacterial template DNA will persist as background in all subsequent steps). The supernatant containing the phage particles can then be transferred to two new reaction tubes, each containing 300 μl ice-cold Phage Precipitation Buffer. After vortexing, phage are precipitated by incubating the vials at room temperature for 10 min, and phage are pelleted by centrifugation for 2 min at maximum speed in a table top centrifuge. Supernatant is removed by pipetting and optional short centrifugation, and both phage pellets are resuspended in 500 μl PBS, and combined. This 1 ml of phage particles in PBS is subjected to a final spin (5 min, table top centrifuge, max speed) to remove aggregated phage and traces of bacteria present. Phage concentration can be determined by measuring the absorbance at 268 nm (An A_{268} of 1.0 corresponds to approximately 1.13×10^{13} phage/ml).

Single stranded uratidyl-DNA can be isolated from the phage particles using the QIAprep spin M13 kit from Qiagen following the manufacturers' instructions (46) (Qiagen, Germany). Visualization of 1 μl of the isolated DNA using a 1% agarose gel, a commercial DNA ladder, the SYBR Safe DNA gel stain and a Safe Imager blue light transilluminator (Invitrogen, USA) will give an indication of the quality of the isolated DNA. The single stranded uritidyl-DNA will run as a single band at approximately one third the molecular weight of your starting plasmid. Higher weight species are usually contaminant M13K07 helper phage DNA. Regular DNA will appear green when stained with SYBR Safe and visualized using the Safe Imager; in contrast the single stranded uratidyl-DNA will appear orange in color. At this point, one has a single-stranded DNA copy of the original phagemid with some of the thymine bases replaced by uridine. One uses this "uracil template" as a template for mutagenesis in the following reactions.

For Kunkel-primer design, similar parameters have to be taken into account as for PCR primer design (47), with the following adaptations. Primers need to be either sense or anti-sense, depending on the directionality of the f1 ori; appropriate overhangs of approximately 20 bases ensure proper annealing. Also, the presence of a guanine or cytosine at the 3′ end of the primer increases primer efficiency. Primers are first phosphorylated: 7 μl primer (at a stock concentration of 0.1 μM) is incubated with 3 μl NEB Ligase Buffer (NEB, USA), 1 μl 10 mM ATP, 1 μl T4 Polynucleotide

Kinase (NEB, USA), and 18 µl deionized water at 37°C for 30 min. After phosphorylation, primers are annealed to the uracil template by directly mixing 1 µl phosphorylated primer with 1 µl NEB Ligase Buffer (NEB, USA) and 1 µl uracil template (add deionized water up to 10 µl total volume). This is repeated for a 10×, a 100× and a 1,000× dilution of primer in deionized water. The molar primer–uracil template ratio has a significant effect on both the number of transformants and the mutagenesis rate (48). For the creation of libraries with degenerate primers, it is advisable to optimize the mutagenesis rate first. As a control, the whole procedure is performed without primers. These mixtures are heated to 95°C and step-wise cooled to 4°C over 5 min using a PCR machine. After cooling, the annealed primers are used in a polymerization reaction: the 10 µl annealing reaction is added to 1.5 µl NEB Ligase Buffer (NEB, USA), 1 µl 100 mM dNTPs, 1 µl 10 mM ATP, 1 µl T7 DNA Polymerase (NEB, USA), 1 µl T4 Ligase (NEB, USA), and 9.5 µl deionized water, mixed and incubated at room temperature for at least 30 min. Kunkel reactions are transformed to an appropriate *E. coli* strain (e.g., XL-1, DH5α).

2.5. Screening

During protein optimization efforts, mutants have to be systematically screened for the desired characteristic(s). Proteins have been selected by Darwinian evolution for millions of years, and are rarely optimized for the conditions in which one wants to apply a biosensor: e.g., specific cell types, organelle-specific parameters such as pH and redox potential, temperature, stability, selectivity, and expression level. Therefore, carefully chosen, stringent parameters need to be set during screening to generate proteins suitable for the target application. For the generation of probes for neuroscience applications, one must always remember that long-term neuronal expression is a particularly demanding setting; testing of intermediate probe designs in the target application is advisable.

3. Development of Genetically Encoded Calcium Indicators

3.1. Overview of GECIs

The secondary messenger calcium (Ca^{2+}) regulates numerous critical physiological processes in a wide range of tissues and organisms, e.g., cellular motility, enzyme activity, and signal transduction. In excitable muscle, cardiac and nerve cells, the influx of calcium ions through voltage-gated channels results in membrane depolarization (49). Both action potentials (APs) and neurotransmitter receptor activation give rise to Ca^{2+} transients, and direct and indirect measurements of calcium can thus be used as a proxy for neuronal spiking and synaptic activity in the brain.

Bioluminescence and fluorescence imaging techniques for visualizing calcium dynamics in living cells and animals using proteins started in 1962 when Johnson and Shimomura described the extraction, purification and properties of the bioluminescent protein aequorin (50). In 1967, Ridgway and Ashley directly injected purified aequorin into single barnacle muscle fibers, and were able to detect calcium transients by bioluminescence (51). It took two more decades before fluorescent protein-based scaffolds for calcium imaging were developed that could compete with the synthetic dyes available; until then bulk and intra-cellular loading of small molecule indicators predominated (52–55). In 1997, two fluorescent protein-based calcium indicators were published, both using calmodulin (CaM) as the calcium-binding protein: FIP-CB$_{SM}$ (56) and Cameleon (57). CaM is a small acidic protein of approximately 150 amino acids, containing four EF-hand motifs, each of which binds one Ca^{2+} ion (58, 59). CaM is an important regulator of cellular function in eukaryotes, and is expressed at high levels in brain and heart tissue (60, 61). CaM undergoes a large conformational change upon calcium binding, during which hydrophobic groups present in the protein become surface exposed. This results in the Ca^{2+}-dependent binding of CaM to hydrophobic protein targets such as the basic amphiphilic helices present in many proteins regulated by CaM (62). One of these targets is the "M13" peptide from myosin light chain kinase (63), used in most CaM-based calcium indicators. FIP-CB$_{SM}$ consists of the M13 peptide sandwiched between blue and green variants of Green Fluorescent Protein (GFP), and relies on endogenous CaM for Ca^{2+} binding. In contrast, Cameleon contains both M13 and CaM in between a blue/green or cyan/yellow fluorescent protein (FP) pair, and was the first fully genetically encoded calcium indicator (GECI) described (FIP-CB$_{SM}$, despite being encodable by DNA, was rather transfected as a protein (56)). Both Cameleon and FIP-CB$_{SM}$ are Förster Resonance Energy Transfer (FRET) (64) sensors, containing two FPs with overlapping excitation and emission spectra (56, 57). Upon calcium binding, the relative distance and orientation of the FPs are altered, resulting in a change in the nonradiative transfer of energy between the donor and acceptor chromophores. This can be detected by excitation of the donor chromophore, and measuring the fluorescence emission of both the donor and acceptor chromophores (for a review of FRET imaging see ref. (65)). Alternatively, the fluorescence of either the donor or acceptor channel may be monitored itself (66).

Several incrementally improved variants of Cameleon have since been published, e.g., the Yellow Cameleons YC2.1, YC2.6, YC3.1, YC4 (67), YC2.3, YC3.3 (68), YC6.1 (69), YC2.12 (70), YC3.6, YC4.6 (71), and VC6.1 (72), the computationally redesigned variants D1cpv (73), D2cpv, D3cpv and D4cpv (74), and the high-affinity YC-Nano (75). (For comparison of FRET-based

Table 2
Comparison of different GCaMP and GCaMP-like GECIs described in literature

Version	In vitro K_d (μM)	Excitation/ emission maximum	$\Delta F/F_0$	pK_a	Brightness compared to EGFP	Reference
GCaMP	0.24	488/510	4.3	7.1	0.002	(82)
GCaMP1.6	0.16	488/509	4.9	8.2	0.09	(83)
GCaMP2	0.16	488/509	5.0	ND	0.53	(84)
Case12	1.0	491/516	12.0	7.2	0.35	(85)
Case16	1.0	490/516	16.5	7.2	0.28	(85)
GCaMP2.1	ND	488/509	5.0	ND	1.47[a]	(86)
GCaMP2.2	ND	498/515	ND	ND	2.57[a]	(86)
GCaMP2.3	ND	ND	ND	ND	2.2[a]	(86)
GCaMP2.4	ND	ND	ND	ND	2.1[a]	(86)
GCaMP3	0.8	498/515	12.0	7.0	2.1[a]	(86)
GCaMP4.1	ND	ND	ND	ND	ND	(87)
GCaMP-HS	0.1	488/509	4.1	ND	ND	(88)

ND not determined
[a]The brightness of these GCaMPs is calculated as relative to that of GCaMP2 in living HEK293 cells

GECIs in cells see also ref. (76).) Additionally, a family of GECIs has been developed based on the muscle-specific Ca^{2+}-binding protein Troponin C: TN-L15 (77), TN-XL (78), and TN-XXL (79).

In 1999, the first single-FP GECI was published: Camgaroo (80). Camgaroo consists of CaM inserted into the yellow fluorescent protein EYFP, near the site of the chromophore. Upon calcium binding, the local environment surrounding the chromophore changes, resulting in an increase in fluorescence. In 2001, two research groups reported variations on this theme: CaM was not inserted into the fluorescent protein, rather the fluorescent protein was circularly permuted at that site, and CaM and the M13 peptide placed at the new termini. These scaffolds were termed Pericam (using YFP) (81) and GCaMP (also "G-CaMP"; using GFP) (82). Since then, several papers have been published on the iterative improvement of the GCaMP scaffold (GCaMP1.6 (83), GCaMP2 (84), Case12 and Case16 (85), GCaMP3 (86), GCaMP4.1 (87), and GCaMP-HS (88)). Crystal structures of several of these proteins have been solved (both in the calcium-bound and -free states), permitting structure-guided engineering of improved variants (89–91). An overview is given in Table 2.

3.2. Case Study: From GCaMP2 to GCaMP3

3.2.1. Goals of Optimization

GECI performance is influenced by many intrinsic and extrinsic parameters (92). Fluorescent indicator response is normally expressed as the signal-to-noise ratio (SNR), which convolves brightness and signal change relative to system-wide error sources. For single-FP GECIs such as GCaMP, we define F_0 as the baseline fluorescence, and F_{obs} as the signal following a calcium-binding event. SNR is expressed as the ratio of the fluorescence change ($\Delta F = F_{obs} - F_0$) to the noise on the baseline fluorescence ($F_0 N^{-1/2}$, where N is the number of photons detected) (93, 94). Probe signal is dictated by many factors: both intrinsic to the sensor (e.g., calcium affinity, dynamic range, brightness, kinetics, specificity, oligomerization state, folding properties) and external contributions (e.g., cytoplasmic calcium concentration at rest and during depolarization events, intracellular calcium buffering effects, and undesired interactions/interference with homeostasis) (for a systematic analysis of the relative contribution of these parameters see ref. (92)). In theory, an optimal GECI would respond quickly to both small and large calcium transients, without perturbing the cell/system overly much. In practice, this situation is difficult to achieve without compromising SNR, dynamic range, or Ca^{2+} buffering; furthermore, extrinsic physiological conditions can vary greatly between cell types and organisms (92). To satisfy demand for GECIs appropriate for a variety of settings and applications, it is likely that a family of sensors will be required. In addition to underlying differences such as temperature and physiology, particular experiments call for indicators with high or low Ca^{2+} affinity (e.g., for detecting sparse activity or extending dynamic range, respectively); bright or dim resting fluorescence (e.g., for localizing labeled cells or extending dynamic range, respectively); fast or slow kinetics (e.g., for resolving spike trains or integrating sparse activity, respectively); desired excitation and emission properties (e.g., color pairing with other labels or indicators, or compatibility with 2-photon excitation for deep, cellular-resolution imaging (95)); and appropriate targeting (e.g., imaging in the soma vs. dendrites (96), using signal peptides (97) or fusion proteins to mediate localization to the pre- or post- synapse (98), among others). Often these parameters must be carefully weighed before choosing GECIs to deploy, and even then it is best to compare several variants in a given application; many aspects of sensor partitioning, buffering, and cytotoxicity are still poorly understood. For a detailed discussion of the trade-offs involved see refs. (92, 99). These different parameters must be taken into account during GECI design and optimization. Here we will explain our strategies to make improved GCaMP variants, with an emphasis on improving SNR, kinetics, and brightness, for the detection of sparse neural activity *in vivo*.

GCaMP2 was the state-of-the-art single wavelength GECI until 2009 (84, 89, 91, 93), but cannot reliably detect single APs in cultured or acute slice (93). The signal-to-noise level is generally significantly smaller *in vivo* than *in vitro* or in slice (motion artifacts, poor optical access, hemodynamics, tissue scattering, etc.). We took it upon ourselves to improve the characteristics of GCaMP2 by both semi-rational design and directed evolution. We will go through the different techniques that resulted in improved variants; references to further information are given when appropriate.

In practice, we follow the following approach during optimization of a genetically encoded calcium indicator: (1) collect structural information on binding proteins and fluorescent proteins used in the biosensor, ideally solving structures of first-generation sensors, (2) target particular regions of the sensor for optimization to improve brightness, fluorescence response (SNR), affinity, kinetics, *in vivo* stability, long-term expression/cytotoxicity etc., (3) create small libraries of variants and subject them to an initial screen (bacterially expressed protein) to remove completely nonfunctional candidates and get a rough group of potentially promising variants (see Sects. 2.4.1 and 3.2.4 (Protocols 1 and 2)), (4) subject interesting variants from the bacterial screen to more stringent rounds of testing by measuring Ca^{2+} transients in human HEK293 cells following the fluorescent response to acetylcholine (Ach)-evoked calcium waves (see Sect. 3.2.6, Protocol 3), (5) test the best variants in neurons by imaging responses to electrical stimulation, and finally, and (6) move the final few best mutants into an *in vivo* setting. Our pipeline thus typically begins (assuming we are improving a sensor with nonzero performance such as GCaMP; the de novo design of new sensor scaffolds will be discussed elsewhere) with the creation of mutants at one or several positions targeted from structural or evolutionary information.

3.2.2. Structure-Guided Optimization of GCaMP

No structural information of GCaMP2 or any other genetically encoded indicator was available at the time we began our studies, therefore we decided to solve the X-ray crystal structure of GCaMP2 (89, 100). To elucidate the sensing mechanism, we crystallized both the calcium-free (apo) and the calcium-bound (sat) states. Much to our surprise, the protein first crystallized as a dimer in the saturated state, in which the CaM domain of one protein was bound to the M13 peptide of the adjacent protein molecule (GCaMP had been reported as a monomer (82)). We verified with size-exclusion chromatography and analytical ultracentrifugation that GCaMP2 did indeed dimerize at high protein concentrations (89). Initial attempts to crystallize calcium-depleted GCaMP2 produced only crystals with Ca^{2+} bound (from buffer, plastic- and glassware contamination). Both these problems were solved using rational selection of point mutants. To hamper crystallization in the dimeric form and improve monomericity for downstream applications, we performed site-directed mutagenesis to target specific residues at the dimer interface (89). To enable crystallization of GCaMP2 without bound calcium, we replaced two calcium-binding amino acids in each of the four EF hands with a glutamine and a glycine. We used gel-filtration chromatography and fluorescence titrations to confirm the disruption of calcium binding and dimerization. With these variants in hand we set out new crystallization trials, and we were able to get structural information for both the calcium-free, and the monomeric, calcium-bound states of the protein (89). GCaMP2 structures were also borne out by the work of other labs (90, 91).

Comparing the calcium-free with the calcium-saturated structure, it became clear how the sensor functions: without calcium, CaM and the M13 peptide are dissociated from each other, resulting in a large hole in the barrel of the circularly permuted GFP (cpGFP) domain. This hole (the "bunghole", analogous to the hole in a wine barrel) can be (partially) plugged by the binding of Ca^{2+}-CaM to the M13 peptide. Upon plugging this bunghole, the local environment of the chromophore in the cpGFP domain changes, resulting in a greater intrinsic fluorescence. Increasing F_{obs} will improve SNR; one way to do this is by "closing the bunghole" even further (ideally reaching the brightness of GFP itself; it might even be possible to exceed this (101)).

We selected a group of amino acids in close proximity to the chromophore in both the apo and bound states for mutagenesis to improve overall GCaMP brightness; we also sought to increase fluorescence of the Ca^{2+}-saturated state by mutating side-chains in the CaM-GFP interface predicted to shield the chromophore upon binding. In parallel, we made small libraries of the Ca^{2+}-binding EF hands and positions known to regulate GFP thermodynamic stability (102). Finally, expression and turn-over of GCaMP was tuned by changing the N-terminally encoded proteasomal degradation sequence (86). All mutants were tested first as bacterially produced protein, and subsequently in HEK293 cells and more intact preparations (86, 89). We discuss our workflow in the following sections.

3.2.3. Bacterial Screening of GECIs

If the system, cell type or condition one uses for screening is very different from the cell type/system one wants to use the protein in, one runs the risk of selecting suboptimal variants ("You get what you screen for") (39). The big trade-off in generating GECIs for neurobiological applications is that it is quite difficult (yet invaluable) to screen for improved variants in neuronal culture or slice preparations, compared to fast and cheap bacteria and HEK293 cells, so a balance has to be struck. To limit the accumulation of false positives, we performed two screens: first an end-point assay in bacterial lysate, and second in live HEK293 cells, measuring a (slow) muscarinic acetylcholine receptor (mAchR)-induced Ca^{2+} transient. The procedures used during our bacterial lysate screen are summarized below in Sect. 3.2.4 (Protocol 2).

3.2.4. Protocol 2: Bacterial
Lysate Screen

Screening of GCaMP2 variants in bacterial lysate was performed using this protocol. For the bacterial-lysate screen, a library or directed mutants, typically generated by Kunkel (45) or QuikChange (Invitrogen, USA) mutagenesis, is transformed to a standard cloning *E. coli* strain (e.g., XL-1 Blue or DH5α). The following day all colonies are scraped from plates, pooled, plasmids isolated, and a small aliquot of this library is subsequently transformed to an *E. coli* expression strain (e.g., BL21(DE3) or Rosetta, EMDBiosciences, USA). First, since we used a leaky bacterial expression vector, pRSET (Invitrogen, USA) lacking the *lacI* gene, we could easily check colony fluorescence (e.g., a microscope or even a simple light source with colored plastic filters such as the SYBR Safe Imager). Variants that appear interesting (fast-maturing, bright, different fluorescent excitation/emission) can in this way be selected for more thorough investigation. Second, colonies are selected at random by hand or using robotics (when using automation such as colony pickers it is important to avoid the largest colonies; they often produce little recombinant protein and therefore are not hampered in growth), and grown in triplicate in 1 ml ZYM-5052 medium (103) containing the appropriate antibiotic for 24–48 h at 30°C in deepwell blocks, shaking. After 24 h, a small aliquot is taken from each well and used to inoculate a fresh 1 ml LB medium containing the appropriate antibiotic. These blocks are incubated at 37°C for at least 16 h, shaking, after which cells can be pelleted and saved for subsequent clone identification by sequencing. External factors such as growth volume, shaking speed, growth time and temperature are very important for the outcome of the screen, and it is advisable to optimize these factors before undertaking a large screen. After sufficient expression, cells are pelleted, the supernatant discarded, and cells lysed to release expressed protein.

For GCaMP mutagenesis, we chose gentle chemical lysis by incubation of the cells in 1 ml lysis buffer (20 mM Tris–HCl pH 8.0, 1 mM $MgCl_2$, 1.5 U DNAse, 1 mg/ml lysozyme), shaking at 30°C for at least 2 h. In this way, cells are fully lysed, and DNA is broken down, which eases liquid handling. Several cycles of freeze-thaw may also improve lysis. Lysate is then clarified by centrifugation (30 min, 4,000×g at 4°C). The fluorescence of 100 μl of clarified lysate is measured as-is, and subsequently apo (2 mM EGTA added) and sat (5 mM $CaCl_2$ final concentration). Besides the fluorescence response of the sensor ($\Delta F/F_0$), these measurements will indicate a rough estimate of F_{apo} and F_{sat} (although brightness measurements can be hard to standardize across experimental procedures, and brightness of the GECI in *E. coli* lysates may not correlate with brightness in model organisms). Factors such as expression level, stability, affinity and maturation are difficult to distinguish in this assay: this must be taken into account during candidate selection. Smaller additions of calcium might reveal high-affinity indicators. Interesting candidates are selected and sequenced using the carbon-copy plates set aside. One can prep the plasmids and directly sequence, however, it is advisable to first PCR the insert using the cells or purified plasmid; DNA quality is often poor coming from *endA⁺ E. coli* expression strains such as BL21(DE3) (104).

Once specific beneficial mutations have been discovered, one has several options to continue the engineering process. First, if structural information is available, one might rationalize the effects of the mutations and propose additional improvements; for example mutagenesis to amino acids with similar characteristics (e.g., glutamate vs. aspartate). Combining mutations can potentially push the sensor to even better performance levels. Second, one might purify small amounts of protein for more detailed assays: e.g., affinity titrations, kinetics measurements, crystallization, etc.

3.2.5. Mammalian Cell Screen

If satisfactory results are obtained one can advance to a more sophisticated GECI assay in living cells. For this, we made use of the intra-cellular release of calcium in response to binding of acetylcholine (Ach) to the muscarinic receptor in HEK293 cells (105). This is an efficient way of GECI screening in a facile mammalian cell screen (86), which sheds light on stability in eukaryotic cells and a crude measure of kinetics (the Ach-induced transients are tens to hundreds of seconds in duration). Addition of decreasing concentrations of Ach to HEK293 cells results in smaller and faster intracellular calcium fluxes, thus providing a more stringent GECI test. In this way, an assessment of different factors such as expression level, basal fluorescence, rise and decay times, affinity, and fluorescence response can be systematically explored in eukaryotic cells. The HEK293 cell screen is not as high-throughput as the bacterial screen, but the sensors are screened more stringently, resulting in fewer false-positives. One example of an indicator performing well in bacterial lysate, but poorly in the HEK293 screen, is the GCaMP2-LIA mutant (86). This variant was picked up during a screen of the linker between the M13 peptide and cpGFP domain from Leucine–Glutamate (the translation of the original *XhoI* site (82)) to three amino acids. The mutant GCaMP2-LIA, with the linker Leucine–Isoleucine–Alanine, showed a two-fold increase in $\Delta F/F_0$ in bacterial lysate and a ~10% increase in $\Delta F/F_0$ in HEK293 cell lysate compared to GCaMP2 (potentially due to the absence of the glutamate which is hydrogen bonded to arginine 81 in the apo state (89), resulting in a lower F_0 for GCaMP2-LIA). However, when GCaMP2-LIA was expressed in HEK293 cells and titrated with Ach, a very small response was seen, and only at high Ach concentrations. Indeed, when expressed in cultured hippocampal slice, GCaMP2-LIA performed significantly worse than GCaMP2 (86), demonstrating the selectivity of the HEK293 cell assay in GECI screening. HEK293 cells, however, lack the calcium extrusion mechanisms and unique geometric considerations of neurons, so this screen should only be considered a crude mechanism to eliminate sensors with inadequately slow responses. For screening of our GCaMP candidates in HEK293 cells, the following protocol was followed.

3.2.6. Protocol 3: HEK293 Screen

Candidates from the bacterial lysate screen that are selected for further analysis are then subcloned into a suitable vector for HEK293 cell expression. For GCaMP screening, we chose not to use a single-shuttle vector system for expression in both bacteria and eukaryotes due to potentially low expression levels in both systems, but rather chose the pEGFP-N1 (Takara-Bio, Japan) vector, driven by a cytomegalovirus (CMV) promoter (106). The CMV immediate early promoter present in this vector produced substantial amounts of GCaMP for imaging, and cells did not appear unhealthy (e.g., no clumping or nuclear filling (86) after transfection and growth. Several different promoters can be used to optimize expression levels if CMV is not strong enough (e.g., CAG, SV40 (107, 108)). HEK293 cells can be transfected in medium-throughput using the 96-well Nucleofector protocol in Amaxa plates (Lonzo, Germany), with each variant in 16 wells for eight duplicate measurements. For GCaMP, cells were incubated at 37°C for 2 days: the incubation time can be tailored to each specific assay and is protein- and promoter-dependent. Every 24 h the growth medium (Dulbecco's Minimum Eagle's Medium with 4.5 g/l glucose (DMEM-HG, Cellgro, USA)) supplemented with Glutamax (Invitrogen, USA) and 10% Fetal Bovine Serum (FBS) (Invitrogen, USA) has to be replaced to ensure optimal cell health. Just before measuring, the cell medium should be aspirated and replaced with 100 µl prewarmed (37 °C) TBS buffer supplemented with 2 mM $CaCl_2$ (Cell-Buffer). A 96-well V-bottom plate containing ~120 µl Cell-Buffer plus Ach in each well was prepared. Acetylcholine should be present in a titration series; a dilution series over eight wells ranging from 10^{-2} to 10^{-10} M is a good start. In this way, each GCaMP variant was subjected to different calcium transients in duplicate. Ligand addition and continuous fluorescence measurement was done for GCaMP simultaneously for all 96 wells using a liquid-handling spectrophotometer equipped with the correct filtersets (e.g., Hamamatsu FDSS). The responses recorded with a liquid handling spectrophotometer will give an indication of resting brightness, the rise time of the indicator, the time it takes for the cells to go back to base-line fluorescence, the level of the response, and at what concentration of acetylcholine a response is visible. These characteristics will give an indication of the potential success of a specific variant, and will weed out sensors only performing well in bacterial lysate. Promising candidates are subsequently tested in neurons or brain slices, following spontaneous or evoked action potentials.

3.2.7. Brain Slice Testing

Imaging GECI responses to action potentials (APs) is the most reliable way to assess relevant sensor kinetics and signal change. After the initial mammalian HEK293 cell screen, potential candidates can be further characterized in acute brain slice using simultaneous two-photon imaging combined with AP stimuli.

For our GCaMP optimization efforts, we used in utero intraventricular electroporation to deliver GECIs to cortical layer2/3 pyramidal neurons (109). To achieve efficient expression, GECIs are subcloned into a pCAG-GS vector, containing a CAG (CMV-enhancer/β-actin) promoter and a regulatory element from the woodchuck hepatitis virus (WPRE) (110). pCAG-mCherry is co-transfected with the GECI (in a molar ratio of ~5:1) for easier visualization and potential normalization of expression. Usually the surgery is performed on E16 mouse embryos. Acute brain slice is then prepared between P14–P17 and cortical layer2/3 pyramidal cells are recorded and imaged simultaneously at room temperature (86).

For electrophysiological recording, defined numbers of APs can be triggered at specific intervals by intermittent short depolarizing pulses (86); alternatively, neurons can be depolarized by continuous current injection from a patch pipette (111). Two-photon calcium imaging of GECI responses to action potential stimuli is typically performed in line-scan mode (500 Hz) across the apical dendrite, 20–50 µm from the base of the neuron (93). A Ti:Sapphire laser (Mai Tai, Spectro-Physics, CA) can be tuned to 910 nm for single-FP green sensors such as GCaMP, and can also be used to excite FRET indicators (e.g., YC-Nano—860 nm to excite CFP (75)). To capture more cells in a field of view, frame scans can be acquired, e.g., 256×256 pixels at 2 Hz for a period of 4 s with a $\times 60$ objective (Olympus, Tokyo, Japan) (86).

To determine GECI response ($\Delta F/F$), the fluorescence baseline is defined as the mean of the 250 ms window immediately prior to AP stimulation, and the peak response is calculated as the maximum value of the filtered trace (100 ms moving window) within 500 ms of stimulation cessation. Compared to the HEK293 cell screen, this method captures many more aspects of *in vivo* functional imaging (e.g., rapid Ca^{2+} flux through voltage-gated channels, light penetration, and possible long-term expression and developmental effects), while maintaining a modest throughput (estimated time is 3 weeks).

3.2.8. In Vivo Applications

Finally, a small number of final mutants are tested in an *in vivo* setting in model organisms (86). Each animal preparation has unique advantages and drawbacks. For GECI development, it can be advantageous to test several organisms at the same time (e.g., *C. elegans*, *Drosophila*, mouse, zebrafish) if possible (86). It is important to remember that GECI performance can vary greatly between target applications (e.g., single AP detection, deconvolution and quantitation of spike trains, large dynamic range over many stimulation regimes, etc.), and it is likely that a family of GECIs will in the end be required to fill every niche of functional calcium imaging.

Standardization of testing during optimization is key, as small changes in conditions might prove to have undesired effects, and will make comparison of variants hard if not impossible. Variables under the user's control typically include: promoter strength, expression method (e.g., electroporation, viral infection, transgenesis), expression time, preimaging preparation (e.g., thinned skull, cranial window, etc.), imaging modality, etc.

4. Notes

In this chapter, we have tried to sketch an overview of the different steps involved in the design, optimization, testing and application of GECIs (see also Fig. 1). We have explained the initial design steps, Kunkel mutagenesis (Sect. 2.4.1, Protocol 1), the bacterial lysate screen (Sect. 3.2.4, Protocol 2), the subsequent HEK293 cell screen (Sect. 3.2.6, Protocol 3), brain slice quantitation and *in vivo* imaging. The methods we described in the different steps draw on diverse skillsets, from protein engineering and structural biology to *in vivo* imaging, and the work is often if not always a team effort.

During GECI design and construction, it is good to select different versions before starting mutagenesis (see Sect. 2.4.1, Protocol 1) (e.g., GECIs containing different calcium binding proteins such as CaM (59) or Troponin C (112), or different fluorescent proteins such as GFP or mWasabi (113, 114)). During optimization (see Sect. 3.2.4, Protocol 2) one always needs to remember the paradigm "you get what you screen for" (39), which implies that characteristics of the GECI not actively screened for (e.g., stability, folding properties, selectivity, affinity, k_{on}/k_{off}) can change during the process, and unwanted factors thus need to be monitored closely. It is tempting to focus on features that can be measured easily (e.g., $\Delta F/F$), resulting in underperforming variants (e.g., GCaMP-LIA). It is therefore essential to test all GECI candidates coming out of the bacterial lysate screen (Sect. 3.2.4, Protocol 2) in the more stringent HEK293 cell screen (Sect. 3.2.6, Protocol 3) (86). It is also important to remember that HEK293 cells are very different from neurons, and promising GECIs performing well in HEK293 cells might not be suited for *in vivo* imaging. It is advisable to test several GECIs coming out of the HEK293 cell screen in brain slice, and to test different promoters and expression tags if necessary (115). Stringent standardized screening with proper parameters has led to improved GECIs (86), and the methods and general procedures described in this chapter are applicable not just to calcium binding proteins but can be used for any fluorescent biosensor scaffold.

Acknowledgments

This work was supported by Howard Hughes Medical Institute, Janelia Farm Research Campus, Ashburn, Virginia, USA. We thank members of the Loogerlab and Fabian Blombach for critical reading of the manuscript.

Note Added in Proof

Two key papers came out since the submission of our methods paper.

1. G-GECO paper by Robert Campbell
   http://www.sciencemag.org/content/333/6051/1888.full
2. GCaMP3 mouse by Tsai-Wen Chen
   http://www.jneurosci.org/content/32/9/3131.long

Furthermore, we have deposited our GCaMP5G construct coming out of this work in addgene: http://www.addgene.org/31788/

References

1. Kazlauskas RJ, Bornscheuer UT (2009) Finding better protein engineering strategies. Nat Chem Biol 5:526–529
2. Marvin SJ, Looger LL (2010) Modulating protein interactions by rational and computational design. In: Sheldon J, Park JRC (eds) Protein engineering and design. CRC Press, Boca Raton, FL, pp 341–364
3. Marshall SA, Lazar GA, Chirino AJ, Desjarlais JR (2003) Rational design and engineering of therapeutic proteins. Drug Discov Today 8:212–221
4. Bloom JD, Arnold FH (2009) In the light of directed evolution: pathways of adaptive protein evolution. Proc Natl Acad Sci USA 106(Suppl 1):9995–10000
5. Scrutton NS, Berry A, Perham RN (1990) Redesign of the coenzyme specificity of a dehydrogenase by protein engineering. Nature 343:38–43
6. Maeda T, Oyama R, Titani K, Sekiguchi K (1993) Engineering of artificial cell-adhesive proteins by grafting EILDVPST sequence derived from fibronectin. J Biochem 113:29–35
7. Gordon DB, Marshall SA, Mayo SL (1999) Energy functions for protein design. Curr Opin Struct Biol 9:509–513
8. Shifman JM, Fromer M (2010) Search algorithms. In: Sheldon J, Park JRC (eds) Protein engineering and design. CRC Press, Boca Raton, FL, p 416
9. Beberg AL, Ensign DL, Jayachandran G, Khaliq S, Pande VS (2009) Folding@home: lessons from eight years of volunteer distributed computing. IEEE Int Symp Parallel Distribut Process 1–5(1624):1631, 3198
10. Kuhlman B, Dantas G, Ireton GC, Varani G, Stoddard BL, Baker D (2003) Design of a novel globular protein fold with atomic-level accuracy. Science 302:1364–1368
11. Jiang L, Althoff EA, Clemente FR, Doyle L, Rothlisberger D, Zanghellini A, Gallaher JL, Betker JL, Tanaka F, Barbas CF 3rd, Hilvert D, Houk KN, Stoddard BL, Baker D (2008) computational design of retro-aldol enzymes. Science 319:1387–1391
12. Siegel JB, Zanghellini A, Lovick HM, Kiss G, Lambert AR, St Clair JL, Gallaher JL, Hilvert D, Gelb MH, Stoddard BL, Houk KN, Michael FE, Baker D (2010) Computational design of an enzyme catalyst for a stereoselective bimolecular Diels-Alder reaction. Science 329:309–313
13. Looger LL, Dwyer MA, Smith JJ, Hellinga HW (2003) Computational design of recep-

tor and sensor proteins with novel functions. Nature 423:185–190

14. Rothlisberger D, Khersonsky O, Wollacott AM, Jiang L, DeChancie J, Betker J, Gallaher JL, Althoff EA, Zanghellini A, Dym O, Albeck S, Houk KN, Tawfik DS, Baker D (2008) Kemp elimination catalysts by computational enzyme design. Nature 453:190–195

15. Schreier B, Stumpp C, Wiesner S, Hocker B (2009) Computational design of ligand binding is not a solved problem. Proc Natl Acad Sci USA 106:18491–18496

16. Khersonsky O, Rothlisberger D, Dym O, Albeck S, Jackson CJ, Baker D, Tawfik DS (2010) Evolutionary optimization of computationally designed enzymes: Kemp eliminases of the KE07 series. J Mol Biol 396: 1025–1042

17. Lutz S (2010) Beyond directed evolution-semi-rational protein engineering and design. Curr Opin Biotechnol 6:734–743

18. Bottomly S, Helmerhorst E (2009) Molecular visualization. In: Bourne P, Gu J (eds) Structural bioinformatics, 2nd edn. John Wiley and Sons, Inc., New Jersey, pp 237–268

19. Lin MZ, McKeown MR, Ng HL, Aguilera TA, Shaner NC, Campbell RE, Adams SR, Gross LA, Ma W, Alber T, Tsien RY (2009) Autofluorescent proteins with excitation in the optical window for intravital imaging in mammals. Chem Biol 16:1169–1179

20. Guntas G, Mansell TJ, Kim JR, Ostermeier M (2005) Directed evolution of protein switches and their application to the creation of ligand-binding proteins. Proc Natl Acad Sci USA 102:11224–11229

21. Dougherty MJ, Arnold FH (2009) Directed evolution: new parts and optimized function. Curr Opin Biotechnol 20:486–491

22. Blagodatski A, Katanaev VL (2010) Technologies of directed protein evolution in vivo. Cell Mol Life Sci 68(7):1207–1214

23. Stemmer WP (1994) Rapid evolution of a protein in vitro by DNA shuffling. Nature 370:389–391

24. Shen B (2002) PCR approaches to DNA mutagenesis and recombination. An overview. Methods Mol Biol 192:167–174

25. Lipovsek D, Pluckthun A (2004) In-vitro protein evolution by ribosome display and mRNA display. J Immunol Methods 290: 51–67

26. Labrou NE (2010) Random mutagenesis methods for in vitro directed enzyme evolution. Curr Protein Pept Sci 11:91–100

27. Chang A, Scheer M, Grote A, Schomburg I, Schomburg D (2009) BRENDA, AMENDA

and FRENDA the enzyme information system: new content and tools in 2009. Nucleic Acids Res 37:D588–D592

28. Cantarel BL, Coutinho PM, Rancurel C, Bernard T, Lombard V, Henrissat B (2009) The Carbohydrate-Active EnZymes database (CAZy): an expert resource for glycogenomics. Nucleic Acids Res 37:D233–D238

29. Consortium TU (2010) The universal protein resource (UniProt) in 2010. Nucleic Acids Res 38:D142–D148

30. Hoffmann R, Valencia A (2004) A gene network for navigating the literature. Nat Genet 36:664

31. Berman H, Henrick K, Nakamura H (2003) Announcing the worldwide Protein Data Bank. Nat Struct Biol 10:980

32. Kelley LA, Sternberg MJ (2009) Protein structure prediction on the Web: a case study using the Phyre server. Nat Protoc 4:363–371

33. Roy A, Kucukural A, Zhang Y (2010) I-TASSER: a unified platform for automated protein structure and function prediction. Nat Protoc 5:725–738

34. Kim DE, Chivian D, Baker D (2004) Protein structure prediction and analysis using the Robetta server. Nucleic Acids Res 32: W526–W531

35. Kuipers RK, Joosten HJ, Verwiel E, Paans S, Akerboom J, van der Oost J, Leferink NG, van Berkel WJ, Vriend G, Schaap PJ (2009) Correlated mutation analyses on super-family alignments reveal functionally important residues. Proteins 76:608–616

36. Davidson AR (2006) Multiple sequence alignment as a guideline for protein engineering strategies. Methods Mol Biol 340: 171–181

37. Kaper T, Brouns SJ, Geerling AC, De Vos WM, Van der Oost J (2002) DNA family shuffling of hyperthermostable beta-glycosidases. Biochem J 368:461–470

38. Brouns SJ, Wu H, Akerboom J, Turnbull AP, de Vos WM, van der Oost J (2005) Engineering a selectable marker for hyperthermophiles. J Biol Chem 280:11422–11431

39. Schmidt-Dannert C, Arnold FH (1999) Directed evolution of industrial enzymes. Trends Biotechnol 17:135–136

40. Stemmer WP (1994) DNA shuffling by random fragmentation and reassembly: in vitro recombination for molecular evolution. Proc Natl Acad Sci USA 91:10747–10751

41. Tracewell CA, Arnold FH (2009) Directed enzyme evolution: climbing fitness peaks one amino acid at a time. Curr Opin Chem Biol 13:3–9

42. Naimuddin M, Kobayashi S, Tsutsui C, Machida M, Nemoto N, Sakai T, Kubo T (2011) Directed evolution of a three-finger neurotoxin by using cDNA display yields antagonists as well as agonists of interleukin-6 receptor signaling. Mol Brain 4:2

43. Kunkel TA, Roberts JD, Zakour RA (1987) Rapid and efficient site-specific mutagenesis without phenotypic selection. Methods Enzymol 154:367–382

44. Wassman CD, Tam PY, Lathrop RH, Weiss GA (2004) Predicting oligonucleotide-directed mutagenesis failures in protein engineering. Nucleic Acids Res 32:6407–6413

45. Kunkel TA, Bebenek K, McClary J (1991) Efficient site-directed mutagenesis using uracil-containing DNA. Methods Enzymol 204:125–139

46. Clackson T, Lowman HB (eds) (2004) Phage display a practical approach. In: Practical approach series no 266. Oxford University Press, Oxford, New York, pp xxiv, p 332

47. Apte A, Daniel S (2009) PCR primer design. Cold Spring Harb Protoc 2009(3): pdb ip65

48. Handa P, Thanedar S, Varshney U (2002) Rapid and reliable site-directed mutagenesis using Kunkel's approach. Methods Mol Biol 182:1–6

49. Yamakage M, Namiki A (2002) Calcium channels – basic aspects of their structure, function and gene encoding; anesthetic action on the channels – a review. Can J Anaesth 49:151–164

50. Shimomura O, Johnson FH, Saiga Y (1962) Extraction, purification and properties of aequorin, a bioluminescent protein from the luminous hydromedusan, Aequorea. J Cell Comp Physiol 59:223–239

51. Ridgway EB, Ashley CC (1967) Calcium transients in single muscle fibers. Biochem Biophys Res Commun 29:229–234

52. Yuste R, Peinado A, Katz LC (1992) Neuronal domains in developing neocortex. Science 257:665–669

53. Svoboda K, Denk W, Kleinfeld D, Tank DW (1997) *In vivo* dendritic calcium dynamics in neocortical pyramidal neurons. Nature 385:161–165

54. Fetcho JR, Cox KJ, O'Malley DM (1998) Monitoring activity in neuronal populations with single-cell resolution in a behaving vertebrate. Histochem J 30:153–167

55. Stosiek C, Garaschuk O, Holthoff K, Konnerth A (2003) *In vivo* two-photon calcium imaging of neuronal networks. Proc Natl Acad Sci USA 100:7319–7324

56. Romoser VA, Hinkle PM, Persechini A (1997) Detection in living cells of Ca^{2+}-dependent changes in the fluorescence emission of an indicator composed of two green fluorescent protein variants linked by a calmodulin-binding sequence – a new class of fluorescent indicators. J Biol Chem 272:13270–13274

57. Miyawaki A, Llopis J, Heim R, McCaffery JM, Adams JA, Ikura M, Tsien RY (1997) Fluorescent indicators for Ca^{2+} based on green fluorescent proteins and calmodulin. Nature 388:882–887

58. Klee CB, Vanaman TC (1982) Calmodulin. Adv Protein Chem 35:213–321

59. Kretsinger RH, Rudnick SE, Weissman LJ (1986) Crystal-structure of calmodulin. J Inorg Biochem 28:289–302

60. Cheung WY (1980) Calmodulin plays a pivotal role in cellular-regulation. Science 207:19–27

61. Cheung WY, Harper JF, Steiner AL, Wallace RW, Wood JG (1980) Calmodulin as a mediator of Ca^{2+} functions. Fed Proc 39:1658

62. Chin D, Means AR (2000) Calmodulin: a prototypical calcium sensor. Trends Cell Biol 10:322–328

63. Blumenthal DK, Takio K, Edelman AM, Charbonneau H, Walsh K, Titani K, Krebs EG (1985) Identification of the calmodulin-binding domain of skeletal-muscle myosin light chain kinase. Biophys J 47:A76

64. Andrews DL (1989) A unified theory of radiative and radiationless molecular-energy transfer. Chem Phys 135:195–201

65. Miyawaki A (2011) Development of probes for cellular functions using fluorescent proteins and fluorescence resonance energy transfer. Annu Rev Biochem 80:327–332

66. Piston DW, Kremers GJ (2007) Fluorescent protein FRET: the good, the bad and the ugly. Trends Biochem Sci 32:407–414

67. Miyawaki A, Griesbeck O, Heim R, Tsien RY (1999) Dynamic and quantitative Ca^{2+} measurements using improved Cameleons. Proc Natl Acad Sci USA 96:2135–2140

68. Griesbeck O, Baird GS, Campbell RE, Zacharias DA, Tsien RY (2001) Reducing the environmental sensitivity of yellow fluorescent protein. Mechanism and applications. J Biol Chem 276:29188–29194

69. Truong K, Sawano A, Mizuno H, Hama H, Tong KI, Mal TK, Miyawaki A, Ikura M (2001) FRET-based *in vivo* Ca^{2+} imaging by a new calmodulin-GFP fusion molecule. Nat Struct Biol 8:1069–1073

70. Nagai T, Ibata K, Park ES, Kubota M, Mikoshiba K, Miyawaki A (2002) A variant of

yellow fluorescent protein with fast and efficient maturation for cell-biological applications. Nat Biotechnol 20:87–90

71. Nagai T, Yamada S, Tominaga T, Ichikawa M, Miyawaki A (2004) Expanded dynamic range of fluorescent indicators for Ca²⁺ by circularly permuted yellow fluorescent proteins. Proc Natl Acad Sci USA 101:10554–10559

72. Evanko DS, Haydon PG (2005) Elimination of environmental sensitivity in a Cameleon FRET-based calcium sensor via replacement of the acceptor with Venus. Cell Calcium 37:341–348

73. Palmer AE, Jin C, Reed JC, Tsien RY (2004) Bcl-2-mediated alterations in endoplasmic reticulum Ca²⁺ analyzed with an improved genetically encoded fluorescent sensor. Proc Natl Acad Sci USA 101:17404–17409

74. Palmer AE, Giacomello M, Kortemme T, Hires SA, Lev-Ram V, Baker D, Tsien RY (2006) Ca²⁺ indicators based on computationally redesigned calmodulin-peptide pairs. Chem Biol 13:521–530

75. Horikawa K, Yamada Y, Matsuda T, Kobayashi K, Hashimoto M, Matsu-ura T, Miyawaki A, Michikawa T, Mikoshiba K, Nagai T (2010) Spontaneous network activity visualized by ultrasensitive Ca²⁺ indicators, yellow Cameleon-Nano. Nat Methods 7:729–732

76. Roe MW, Fiekers JF, Philipson LH, Bindokas VP (2006) Visualizing calcium signaling in cells by digitized wide-field and confocal fluorescent microscopy. Methods Mol Biol 319:37–66

77. Heim N, Griesbeck O (2004) Genetically encoded indicators of cellular calcium dynamics based on troponin C and green fluorescent protein. J Biol Chem 279:14280–14286

78. Mank M, Reiff DF, Heim N, Friedrich MW, Borst A, Griesbeck O (2006) A FRET-based calcium biosensor with fast signal kinetics and high fluorescence change. Biophys J 90: 1790–1796

79. Mank M, Santos AF, Direnberger S, Mrsic-Flogel TD, Hofer SB, Stein V, Hendel T, Reiff DF, Levelt C, Borst A, Bonhoeffer T, Hubener M, Griesbeck O (2008) A genetically encoded calcium indicator for chronic in vivo two-photon imaging. Nat Methods 5:805–811

80. Baird GS, Zacharias DA, Tsien RY (1999) Circular permutation and receptor insertion within green fluorescent proteins. Proc Natl Acad Sci USA 96:11241–11246

81. Nagai T, Sawano A, Park ES, Miyawaki A (2001) Circularly permuted green fluorescent proteins engineered to sense Ca²⁺. Proc Natl Acad Sci USA 98:3197–3202

82. Nakai J, Ohkura M, Imoto K (2001) A high signal-to-noise Ca²⁺ probe composed of a single green fluorescent protein. Nat Biotechnol 19:137–141

83. Ohkura M, Matsuzaki M, Kasai H, Imoto K, Nakai J (2005) Genetically encoded bright Ca²⁺ probe applicable for dynamic Ca²⁺ imaging of dendritic spines. Anal Chem 77:5861–5869

84. Tallini YN, Ohkura M, Choi BR, Ji G, Imoto K, Doran R, Lee J, Plan P, Wilson J, Xin HB, Sanbe A, Gulick J, Mathai J, Robbins J, Salama G, Nakai J, Kotlikoff MI (2006) Imaging cellular signals in the heart in vivo: cardiac expression of the high-signal Ca²⁺ indicator GCaMP2. Proc Natl Acad Sci USA 103:4753–4758

85. Souslova EA, Belousov VV, Lock JG, Stromblad S, Kasparov S, Bolshakov AP, Pinelis VG, Labas YA, Lukyanov S, Mayr LM, Chudakov DM (2007) Single fluorescent protein-based Ca²⁺ sensors with increased dynamic range. BMC Biotechnol 7:37

86. Tian L, Hires SA, Mao T, Huber D, Chiappe ME, Chalasani SH, Petreanu L, Akerboom J, McKinney SA, Schreiter ER, Bargmann CI, Jayaraman V, Svoboda K, Looger LL (2009) Imaging neural activity in worms, flies and mice with improved GCaMP calcium indicators. Nat Methods 6:875–881

87. Shindo A, Hara Y, Yamamoto TS, Ohkura M, Nakai J, Ueno N (2010) Tissue-tissue interaction-triggered calcium elevation is required for cell polarization during Xenopus gastrulation. PLoS One 5:e8897

88. Muto A, Ohkura M, Kotani T, Higashijima SI, Nakai J, Kawakami K (2011) Genetic visualization with an improved GCaMP calcium indicator reveals spatiotemporal activation of the spinal motor neurons in zebrafish. Proc Natl Acad Sci USA 108(13):5425–5430

89. Akerboom J, Rivera JD, Guilbe MM, Malave EC, Hernandez HH, Tian L, Hires SA, Marvin JS, Looger LL, Schreiter ER (2009) Crystal structures of the GCaMP calcium sensor reveal the mechanism of fluorescence signal change and aid rational design. J Biol Chem 284:6455–6464

90. Leder L, Stark W, Freuler F, Marsh M, Meyerhofer M, Stettler T, Mayr LM, Britanova OV, Strukova LA, Chudakov DM, Souslova EA (2010) The structure of Ca²⁺ sensor Case16 reveals the mechanism of reaction to low Ca²⁺ concentrations. Sensors 10: 8143–8160

91. Wang Q, Shui B, Kotlikoff MI, Sondermann H (2008) Structural basis for calcium sensing by GCaMP2. Structure 16:1817–1827

92. Hires SA, Tian L, Looger LL (2008) Reporting neural activity with genetically encoded calcium indicators. Brain Cell Biol 36:69–86

93. Mao T, O'Connor DH, Scheuss V, Nakai J, Svoboda K (2008) Characterization and sub-cellular targeting of GCaMP-type genetically-encoded calcium indicators. PLoS One 3: e1796

94. Yasuda R, Nimchinsky EA, Scheuss V, Pologruto TA, Oertner TG, Sabatini BL, Svoboda K (2004) Imaging calcium concentration dynamics in small neuronal compartments. Sci STKE 2004(219):pl5

95. Kerr JN, Denk W (2008) Imaging *in vivo*: watching the brain in action. Nat Rev Neurosci 9:195–205

96. Borghuis BG, Tian L, Xu Y, Nikonov SS, Vardi N, Zemelman BV, Looger LL (2011) Imaging light responses of targeted neuron populations in the rodent retina. J Neurosci 31:2855–2867

97. McCombs JE, Palmer AE (2008) Measuring calcium dynamics in living cells with genetically encodable calcium indicators. Methods 46:152–159

98. Dreosti E, Odermatt B, Dorostkar MM, Lagnado L (2009) A genetically encoded reporter of synaptic activity *in vivo*. Nat Methods 6:883–889

99. Tian L, Looger LL (2008) Genetically encoded fluorescent sensors for studying healthy and diseased nervous systems. Drug Discov Today Dis Model 5:27–35

100. Rodriguez Guilbe MM, Alfaro Malave EC, Akerboom J, Marvin JS, Looger LL, Schreiter ER (2008) Crystallization and preliminary X-ray characterization of the genetically encoded fluorescent calcium indicator protein GCaMP2. Acta Crystallogr Sect F Struct Biol Cryst Commun 64:629–631

101. Kirchhofer A, Helma J, Schmidthals K, Frauer C, Cui S, Karcher A, Pellis M, Muyldermans S, Casas-Delucchi CS, Cardoso MC, Leonhardt H, Hopfner KP, Rothbauer U (2010) Modulation of protein properties in living cells using nanobodies. Nat Struct Mol Biol 17:133–138

102. Pedelacq JD, Cabantous S, Tran T, Terwilliger TC, Waldo GS (2006) Engineering and characterization of a superfolder green fluorescent protein. Nat Biotechnol 24:79–88

103. Studier FW (2005) Protein production by auto-induction in high density shaking cultures. Protein Expr Purif 41:207–234

104. Taylor RG, Walker DC, McInnes RR (1993) *E. coli* host strains significantly affect the quality of small scale plasmid DNA preparations used for sequencing. Nucleic Acids Res 21:1677–1678

105. Birdsall NJ, Hulme EC, Keen M, Pedder EK, Poyner D, Stockton JM, Wheatley M (1986) Soluble and membrane-bound muscarinic acetylcholine receptors. Biochem Soc Symp 52:23–32

106. Foecking MK, Hofstetter H (1986) Powerful and versatile enhancer-promoter unit for mammalian expression vectors. Gene 45: 101–105

107. Yew NS (2005) Controlling the kinetics of transgene expression by plasmid design. Adv Drug Deliv Rev 57:769–780

108. Alexopoulou AN, Couchman JR, Whiteford JR (2008) The CMV early enhancer/chicken beta actin (CAG) promoter can be used to drive transgene expression during the differentiation of murine embryonic stem cells into vascular progenitors. BMC Cell Biol 9:2

109. Walantus W, Castaneda D, Elias L, Kriegstein A (2007) *In utero* intraventricular injection and electroporation of E15 mouse embryos. J Vis Exp 6:e239

110. Niwa H, Yamamura K, Miyazaki J (1991) Efficient selection for high-expression transfectants with a novel eukaryotic vector. Gene 108:193–199

111. Hendel T, Mank M, Schnell B, Griesbeck O, Borst A, Reiff DF (2008) Fluorescence changes of genetic calcium indicators and OGB-1 correlated with neural activity and calcium *in vivo* and *in vitro*. J Neurosci 28: 7399–7411

112. Ebashi S (1963) Third component participating in the superprecipitation of 'natural actomyosin'. Nature 200:1010

113. Tsien RY (1998) The green fluorescent protein. Annu Rev Biochem 67:509–544

114. Ai HW, Olenych SG, Wong P, Davidson MW, Campbell RE (2008) Hue-shifted monomeric variants of Clavularia cyan fluorescent protein: identification of the molecular determinants of color and applications in fluorescence imaging. BMC Biol 6:13

115. Lorkowski S, Cullen P (2003) Analysing gene expression: a handbook of methods: possibilities and pitfalls. Wiley-VCH, Weinheim; New York

116. Erlich HA (1989) PCR technology: principles and applications for DNA amplification. Macmillan Publishers; New York, NY

117. Liu R, Barrick JE, Szostak JW, Roberts RW (2000) Optimized synthesis of RNA-protein fusions for in vitro protein selection. Methods Enzymol 318:268–293

118. Gold L (2001) mRNA display: diversity matters during in vitro selection. Proc Natl Acad Sci USA 98:4825–4826

119. Smith GP (1985) Filamentous fusion phage: novel expression vectors that display cloned antigens on the virion surface. Science 228: 1315–1317

120. Barbas CF (2001) Phage display: a laboratory manual. Cold Spring Harbor Laboratory Press, Cold Spring Harbor, N.Y

121. Kay BK, Winter J, McCafferty J (1996) Phage display of peptides and proteins: a laboratory manual. Academic Press, San Diego

Chapter 9

Imaging cAMP Dynamics in the Drosophila Brain with the Genetically Encoded Sensor Epac1-Camps

Katherine R. Lelito and Orie T. Shafer

Abstract

Cyclic adenosine monophosphate (cAMP) is a critical second messenger signaling molecule, the modulation of which is a central mode of nervous system function. The development of genetically encoded sensors for cAMP has made it possible to measure cAMP dynamics within living cells with high temporal and spatial resolution. In this chapter, we review live-imaging methods for the measurement of relative cAMP levels within single neurons of the explanted adult *Drosophila* brain. Specifically, we describe a detailed protocol for the monitoring of neuronal cAMP dynamics during bath application of neurotransmitters and neuromodulators using the genetically encoded cAMP sensor Epac1-camps and scanning laser confocal microscopy.

Key words: cAMP, FRET, Epac1-camps, Live-imaging, *Drosophila*, Neurons, Brain, Neurotransmitter, Neuromodulator, Bath-application, Second messenger, Sensor

1. Introduction

Despite the relative simplicity of its nervous system, *Drosophila melanogaster* is capable of producing a remarkable repertoire of complex behaviors (1). This, coupled with its remarkable genetic malleability, has made the fly a valuable model system for the molecular and genetic basis of animal behavior. The discovery of genes that govern specific behaviors in *Drosophila* has allowed neurobiologists to identify the putative neural substrates of those behaviors through the mapping of behavioral gene expression within the central nervous system (2). However, the electrophysiological inaccessibility of many of the fly's central brain neurons presents a fundamental challenge for the *Drosophila* neurobiologist. Understanding how networks of genetically defined neurons produce specific behavioral outputs requires the ability to measure

Jean-René Martin (ed.), *Genetically Encoded Functional Indicators*, Neuromethods, vol. 72,
DOI 10.1007/978-1-62703-014-4_9, © Springer Science+Business Media, LLC 2012

the physiological responses of these networks to sensory and neurochemical stimuli. In this regard, the development of genetically encoded sensors for neuronal activity represents a critical technical development (3).

The creation of green fluorescent protein (GFP)-based Ca^{2+} sensors whose expression can be driven within genetically defined neural networks of the fly has made it possible to measure the activity of neuronal populations in response to sensory input and neurochemical modulation (4). The optimization of GCaMP, a GFP-based Ca^{2+} sensor, has produced a robust and sensitive sensor for the detection of neuronal excitation in the fly brain (5). Nevertheless, many important neuromodulators act through second messenger signaling cascades that may have little or no acute effects on Ca^{2+}. For example, cyclic adenosine monophosphate (cAMP) is a ubiquitous signaling molecule, the modulation of which underlies many important physiological processes (6–8). In both flies and mammals cAMP signaling is critical for the formation of long-term memories (9), the transduction of olfactory cues (10, 11), and the maintenance of endogenous circadian rhythms (12, 13). The ability to measure cAMP dynamics within neural networks of *Drosophila* would allow for the measurement of behaviorally relevant neuromodulatory cascades that might not be easily detected using Ca^{2+} sensors.

Recently, several genetically encoded sensors for cAMP have been developed (reviewed in ref. (14)). Two of these, GFP-PKA and Epac1-camps, have been successfully used to measure relative cAMP levels within neural networks of the fly brain with high spatial and temporal resolution (15, 16). Another sensor designed to specifically measure protein kinase A (PKA) activation has also been successfully employed in the fly brain (17). In this chapter, we describe methods for measuring relative cAMP levels in single *Drosophila* neurons in the explanted brain using the Epac1-camps sensor and scanning laser confocal microscopy. We direct the reader to the excellent protocol of Börner et al. (18) for the measurement of cAMP concentrations in cell culture using Epac1-camps and epifluourescent microscopy.

Epac1-camps was created in the laboratory of Martin Lohse, one of several labs to develop a sensor based on Epac (exchange protein activated by cAMP) (19–21). Epac is a guanine nucleotide exchange factor of Rap1, a Ras-like GTPase that mediates PKA-independent effects of cAMP (22). The Epac-based cAMP sensors all rely on Förster resonance energy transfer (FRET) for the detection of changing cAMP levels. These sensors consist of Epac sequences flanked by the GFP variants cyan fluorescent protein (CFP) and yellow fluorescent protein (YFP) (19–21). When unbound to cAMP the proximity of CFP and YFP allows for energy transfer from CFP to YFP so that the excitation of CFP with 440 nm light results in relatively high YPF emission and low CFP emission in the

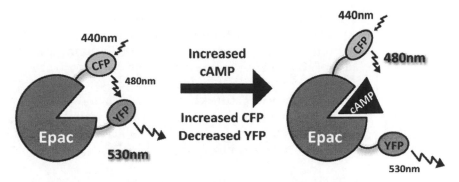

Fig. 1. The genetically encoded cAMP sensor Epac1-camps: the Epac1-camps is derived from the cAMP binding domain of Epac, a guanine nucleotide exchange factor of Rap1, a Ras-like GTPase. This Epac domain is flanked by cyan fluorescent protein (CFP) and yellow fluorescent protein (YFP) and reports cAMP changes through fluorescence resonance energy transfer (FRET) between CFP and YFP. When unbound to cAMP, the proximity of CFP and YFP allows for energy transfer from CFP to YFP so that the excitation of CFP with 440 nm light results in relatively high YPF emission at 530 nm and low CFP emission and 480 nm. Binding of a single molecule of cAMP results in a conformational change of the Epac1-camps sensor that increases the distance between CFP and YFP, thereby reducing energy transfer. Epac1-camps therefore displays concomitant increases in CFP and decreases in YFP emission upon increased cAMP levels. Thus, the FRET ratio (YFP/CFP) of Epac-based sensors is inversely proportional to cAMP levels (19).

absence of bound cAMP. Upon the binding of a single molecule of cAMP, Epac-based sensors undergo a conformational change that increases the distance between CFP and YFP, thereby reducing energy transfer. Thus, when cAMP levels rise, Epac-based sensors display concomitant increases in CFP and decreases in YFP emission. In this manner, the FRET ratio (YFP/CFP) of Epac-based sensors is inversely proportional to cAMP levels. Figure 1 schematizes the general structure and function of Epac-based cAMP sensors.

Unlike the PKA-based cAMP sensor GFP-PKA, Epac1-camps is well tolerated by fly neurons (15, 16). This is likely a reflection of the fact that GFP-PKA contains functional domains of PKA that may perturb cAMP/PKA signaling while Epac1-camps has been stripped of Epac's catalytic domains (discussed in refs. (14, 23)). Furthermore, the invariant 1:1 CFP to YFP stoichiometry of the single molecule Epac1-camps makes its measurement of relative cAMP levels more straightforward than that of the GFP-PKA sensor, which consists of separate CFP-bound regulatory and YFP-bound catalytic subunits of PKA (19, 24, 25). Several laboratories have used Epac1-camps to measure cAMP dynamics within the neural networks of the fly that govern circadian rhythms (16, 26), olfactory learning (27), and sleep (28, 29). It is important to note that despite that fact that Epac1-camps is nearly ten-times more sensitive to cAMP than to cyclic guanosine monophosphate (cGMP) in vitro (19), it appears to be sensitive to physiologically relevant cGMP levels in fly neurons (30). Therefore, when using Epac1-camps, conclusions regarding cAMP must be tempered by the possibility that changes in cGMP have caused or contributed to observed changes in Epac1-camps FRET.

In this chapter we present a detailed protocol for using Epac1-camps to measure relative cAMP levels within single neurons of the explanted *Drosophila* brain during the administration of bath-applied neurotransmitters. We describe the use of the GAL4/UAS binary system to drive Epac1-camps expression within defined neuronal networks, the dissection and preparation of the brain for imaging, the collection of time-course imaging data, the delivery of neurochemical stimuli, and the processing and analysis of Epac1-camps time-course data.

2. Materials

2.1. Fly Crosses and Rearing

Protocols for the rearing of *Drosophila* in the laboratory have been described in detail by Ashburner and Roote (31) and are not discussed in detail here. Three *UAS-Epac1-camps* lines are available as independent insertions in chromosomes one, two, and three from the Bloomington Drosophila Stock Center (http://flystocks. bio.indiana.edu/) and GAL4 driver lines are available from labs and stock centers around the world. The reader is directed to the appendix of Greenspan's introduction of *Drosophila* genetics for stock center information (32). A simple cross between a *UAS-Epac1-camps* line and an appropriate GAL4 line will yield offspring with UAS-Epac1-camps expression in GAL4 expressing neurons (see below). We recommend using brains from 3- to 7-day-old adult flies for live imaging experiments.

2.2. Dissection and Mounting

Adult brain dissection requires a dissecting stereomicroscope, a sylgard dissecting dish, an insect pin, two sets of sharp fine forceps and ice-cold fly Ringer's solution consisting of (in mM): 128 NaCl, 2 KCl, 1.8 CaCl2, 4 $MgCl_2$, 35.5 sucrose, 5 HEPES, pH 7.1 (33). Fly saline can be autoclaved and stored at room temperature. After complete dissection of the brain, it should be transferred to the bottom of a 35 mm plastic culture dish containing the hemolymph-like saline HL3 consisting of (in mM): 70 NaCl, 5 KCl, 1.5 CaCl2, 20 $MgCl_2$, 10 NaHCO3, 5 D(+) Trehelose, 115 sucrose, 5 HEPES, pH 7.1 (34). HL3 should be vacuum filtered and frozen in small (~50 ml) aliquots. We recommend using freshly thawed HL3 for each day's experiments.

2.3. Recording and Data Analysis

This protocol requires a scanning laser confocal microscope with its associated command computer and software. The confocal must be capable of scanning optical sections using a laser with a wavelength at or near 440 nm and separating CFP and YFP emission centered around 480 and 530 nm, respectively. The microscope should also be equipped with epifluorescence optics for CFP and GFP or YFP visualization. The precise nature of the lenses, laser lines, and dichroic mirrors used for FRET time-course experiments will vary among imaging systems. Details of time-course acquisition, data collection and analysis will also vary among acquisition and analysis programs.

2.4. Solutions and Bath-Application

HL3 saline and dissolved compounds can be administered using a gravity fed-perfusion system. These systems typically comprise a series of elevated reservoirs fed via small gauge plastic tubes to a rubber chamber that directs fluid flow across the preparation and make possible a rapid switching between reservoirs. Fluid is removed from the dish via peristaltic or vacuum pumping. Alternatively, compounds can be prepared as 10× solutions and directly micropipetted into a predetermined volume of HL3 within the dish to create a 1× final concentration of compound. This is the suggested method for compounds that cannot be perfused at high volumes due to scarcity or expense. Forskolin can be prepared as a 10 mM stock solution in DMSO and frozen in small (10–100 µl) aliquots. These aliquots can be freshly diluted in HL3 for imaging experiments. We recommend dissolving neurotransmitters, agonists, and antagonists in freshly made or freshly thawed HL3 immediately before conducting live imaging experiments. Researchers should strive to minimize the amount of time a compound sits in aqueous solution before being applied to brains.

3. Methods

The procedures we describe in this chapter are designed specifically for monitoring cAMP dynamics within neurons of the explanted *Drosophila* brain in response to bath-application of neurochemical stimuli. Nevertheless, many aspects of this protocol may be applied to other tissues, organisms, and types of stimuli.

3.1. Tissue Preparation

In *Drosophila*, the targeted expression of genetically encoded sensors is most easily achieved using the GAL4/UAS system (35). In this system the yeast transcription factor GAL4 is driven by endogenous fly enhancers, so that GAL4 is expressed in a manner that resembles the expression of the fly gene controlled by the enhancer in question. Transgenes, such as genetically encoded sensors, are placed downstream of the upstream activating sequence (UAS), GAL4's target promoter sequence in yeast. The generation of flies expressing the *UAS-Epac1-camps* under the control of GAL4 elements is the most time consuming aspect of this protocol, taking about 2 weeks for adult *GAL4/UAS-Epac1-camps* flies to develop. Such flies can be easily created by crossing virgin *UAS-Epac1-camps* flies to male flies carrying a GAL4 element that is expressed in neurons of interest (Fig. 2a). We direct the reader

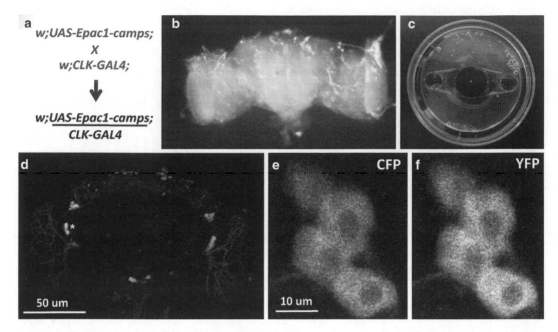

Fig. 2. Imaging Epac1-camps expression in the *Drosophila* brain for cAMP time-course experiments. (**a**) A simple cross scheme for the expression of *UAS-Epac1-camps* within the clock neuron network of the adult fly. (**b**) A dissected brain adhered to the bottom of a culture dish under HL3 saline. (**c**) An adult *Drosophila* brain mounted in a 35 mm culture dish within a perfusion insert. The brain is the *white dot* in the center of the large middle hole. The small left and right holes are the input and output of the perfusion line. (**d**) A projected confocal Z-series reconstruction of Epac1-camps expression in the clock neurons of an adult *w;UAS-Epac1camps/Clock-GAL4* brain. An *asterisk* indicates the cluster of large ventrolateral neurons (l-LNv) in the left hemisphere. (**e**) A single optical section of CFP expression through a cluster of l-LNv of an adult *w;UAS-Epac1camps/Clock-GAL4* brain. (**f**) YFP expression from the same optical section as shown in (**e**). When scanned with a 440 nm laser, the ratio of YFP/CFP emission intensities is inversely proportional to cAMP levels.

unfamiliar with fly genetics to Greenspan's excellent introduction to *Drosophila* (32). Several *UAS-Epac1-camps* lines were created in the laboratory of Paul Taghert (16), and are available through the Bloomington Drosophila Stock Center at Indiana University (see above). Hundreds of GAL4 lines are available from stock centers and laboratories around the world. The GAL4 element selected should allow for reliable identification of the neuron classes of interest across individuals. For example, we have used a Clock-GAL4 element (36) to drive UAS-Epac1-camps expression specifically in the clock neuron network of the adult fly brain. Using the most specific GAL4 line possible will minimize the chances of recording Epac1-camps FRET from nontarget neurons and will make the identification of neurons of interest relatively easy. Nevertheless, given the high Z-resolution of the confocal microscope, even widely expressed GAL4 elements can be used if the experimenter is able to reliably identify and optically isolate the target neurons of interest within the brain. It is possible to create lines of flies that stably maintain both GAL4 and UAS-Epac1-camps elements using standard fly genetic methods (32), thereby obviating the need to set up crosses for each experiment (but see note 1).

Three to five days after the GAL4/UAS-bearing flies emerge, brains may be dissected and used for live imaging experiments. To dissect brains, immobilize adult flies on ice. Pin individual flies down under ice-cold Ringer's solution within a sylgard lined dissecting dish using a single insect pin through the thorax. Carefully disrupt the head cuticle using two fine forceps and gently remove all cuticle and (if desired) compound eye tissue from the underlying brain (Fig. 2b). A detailed protocol for *Drosophila* brain dissection and a description of the necessary materials has been described in detail (37). Once isolated, the brain can be adhered to the bottom of a clean 35 mm culture dish beneath a large drop of HL3. In our experience, the fly brain adheres sufficiently to plastic culture dishes, eliminating the need for poly-lysine or other treatments (Fig. 9.2b, c; see note 2). When immobilizing flies on ice for dissections, it is important to never use brains from flies that have been on ice for more than an hour. Dissected brains can be kept alive in HL3 for several hours, though we caution the reader from extending experiments beyond an hour following dissection and mounting. For experiments using a perfusion system, a perfusion insert can be lowered around the brain (Fig. 2c). For simple pipette-based application, the brain can simply be submerged in a predetermined volume of HL3.

It is critical to have an intact brain for Epac1-camps imaging experiments and it is important that only cleanly dissected brains that have not been punctured or otherwise disrupted be used for experiments. It is therefore critical that the mounted brain be carefully inspected for disruptions and that disrupted brains be discarded. After a cleanly dissected brain is mounted, it should be allowed to stabilize for 10–20 min before the experiment is conducted. This will prevent the experimenter from recording fluctuations in cAMP levels that are a direct response to the trauma of the dissection. This wait also allows the brain to settle onto the culture dish, which will minimize settling movements during imaging (see note 3).

In our experience, moderate to moderately high expression levels of the Epac1-camps sensor are needed to reliably measure the changes in Epac1-camps FRET associated with changes in cAMP. However, extremely high Epac1-camps expression can be problematic (18) and it is critical that positive controls are used to insure that cAMP dependent FRET changes can be measured for each neuron class and GAL4/UAS combination used (see below). Expression levels will depend on the strength of GAL4 expression by the cell type in question (see note 4). The level of Epac1-camps expression can be determined visually by observing the YPF fluorescence under epifluorescent illumination using standard YFP or GFP optics. Expression levels are typically sufficient when the

Epac1-camps expressing neurons are easily visible under epifluorescent illumination and when relatively low laser powers are required to visualize CFP and YFP emission using the scanning laser confocal (Fig. 2d–f). The absolute values of the laser power will vary among imaging systems, but should be low enough as to not cause a rapid photobleaching of CFP or YFP.

3.2. Recording Parameters

Optics for CFP/YFP FRET should be used for Epac1-camps imaging experiments, whereby the tissue is scanned with a 440 nm laser and CFP and YFP emission wavelengths are collected at 480 and 530 nm, respectively (24). Epac1-camps-expressing cell bodies and neurites can typically be imaged using a ×20 or higher magnification objective. For upright microscopes, dipping cone objectives should be used. For inverted microscopes, standard objectives can be used, but in this case culture dishes with coverglass bottoms should be employed for optimal image quality. In this case the bottoms of the dishes should be coated with poly-lysine to insure brain adherence. Neurons of interest can be located first using epifluorescent illumination followed by rapid laser scanning and fine focal adjustment. Upon determining the focal plane and imaging field for a particular set of neurons, regions of interest (ROIs) can be placed over the cell bodies or neurites contained in the imaging plane for collection of fluorescent intensity values over time. This is easily accomplished using the software controlling most imaging systems.

Changes in cAMP generally occur over timescales of tens of seconds to several minutes. Therefore, scanning frequencies of 0.1–1 Hz are usually sufficient to measure the kinetics of cAMP responses to bath-applied neurotransmitter in the fly brain (16, 27). Sampling at higher frequencies will typically reveal very little and can result in photobleaching of Epac1-camp's fluorophores. YFP is usually more sensitive to such photobleaching, which results in an apparent decrease in the YFP/CFP FRET ratio (18). Therefore, to ensure consistency of baseline Epac1-Camps FRET profiles between samples, laser power, dwell time, and scanning frequency should remain constant between the samples of a given experiment (see note 5). We favor a strategy in which these settings are optimized for a given GAL4/UAS combination and neuron type based on several 10-min forskolin and vehicle control time-course experiments (see below). Settings should make possible the detection of FRET changes in response to the former but should not result in significant changes in FRET for the latter. Once these settings have been determined, they may be used repeatedly for specific bath-application experiments. Note: these settings may differ among cell types, even when the same GAL4/UAS combination is used, due to variation in GAL4/UAS expression and the different depths in which the cell types reside.

It is best to determine the length of the recording time for a particular experiment empirically by testing response latencies during trial experiments. In our experience, response and recovery times range between 1 and 10 min, depending on the nature of transmitters and receptors underlying the response. Ionotropic responses are often relatively rapid (tens of seconds) while metabotropic responses can take many minutes (see note 6). Thus, 10-min time-courses will be sufficient for most bath-application experiments, though shorter experiments will suffice for some responses. Several factors likely influence the latency of an Epac1-camps FRET response. For example, bath-applied neurotransmitter must diffuse through the neurolemma into the brain and various transmitters likely differ in their permeability and diffusion rates. The application of neurotransmitters via directional perfusion or picosprtizer puffing minimizes variability in response latency. Variability in dissection quality may also lead to differential diffusion rates of compounds through the tissues of different brains. Thus, it is important to strive for exact and consistent dissections when conducting such live imaging experiments.

3.3. Applying Stimuli

To determine the effects of a bath-applied compound on the cAMP dynamics of a particular neuron type, changes in Epac1-camps FRET can be recorded before, during, and after the controlled application of the compound. We favor the use of a perfusion system to deliver compounds to the dissected brain. Perfusion flow allows for the precise timing of application, for rapid removal of the compound, and for the application of several test compounds in series. Alternatively, for compounds that cannot be perfused in large volumes, such as synthesized peptides or expensive compounds, a concentrated aliquot can be directly applied to the bath during the recording using a micropipette (16). Alternatively, discrete regions of the brain can be stimulated using a picosprtizer positioned directly over a brain area of interest (27). Each of these methods of application can cause movement in the specimen that can cause detectable changes in FRET values (e.g., Fig. 3). These movement artifacts should be minimized through the adjustment of stimulation parameters during initial trials of vehicle administration. Nevertheless, some level of movement artifact will likely be unavoidable and the inclusion of vehicle controls, the filtering of time-course data (see below), and the ratiometric nature of Epac1-camps measurement can all serve to minimize the impact of such artifacts on the final analysis of cAMP dynamics (see note 3).

The establishment of a stable FRET baseline is necessary before the effects of a bath-applied compound can be accessed. When cAMP levels are stable, the baseline FRET ratio should be relatively flat, though some drift in the ratio due to slow YPF bleaching is sometimes unavoidable. If drastic ratio fluctuations are occurring in the moments leading to compound application it is difficult, if not impossible, to determine the compound's effects on cAMP dynamics (see note 7). In such cases, the specimen may require

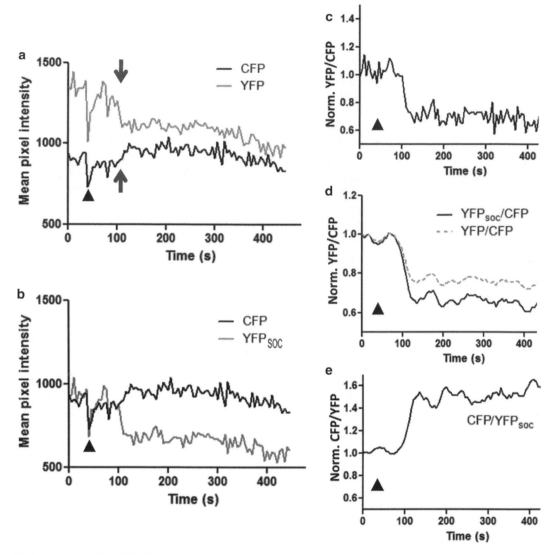

Fig. 3. Data processing of FRET responses from Epac1-camps. (**a**) CFP and YFP emission from a single large ventrolateral clock neuron within a *w;UAS-Epac1camps/Clock-GAL4* brain before, during, and after the addition of pipetted forskolin (final concentration 20 μM) . CFP emission is plotted in *black* and YFP in *grey*. The application of forskolin is indicated by the *black triangle*. The *arrows* indicate the simultaneous decrease in YFP and increase in CFP associated with true FRET responses. (**b**) The same data as for (**a**) but using spillover-corrected YFP values. (**c**) The raw spillover corrected (SOC) YFP/CFP ratio from the data in (**b**). (**d**) Normalized FRET ratio plots that have been filtered using a five-point moving average. The ratio values using SOC YFP are plotted in *black*. Broken *grey lines* represent the uncorrected values. (**e**) Inversion of the normalized Epac1-camps FRET ratio makes the ratio change proportional to relative cAMP levels.

more time to equilibrate or may be damaged and therefore unsuitable for imaging. Nevertheless, some fly neuron classes do appear to support rhythmic cAMP fluctuations and it is possible to measure cAMP responses despite such fluctuations (16).

It is possible to record multiple time-courses from the same neurons in response to a series of treatments. In these cases, the complete removal of compounds must be ensured before subsequent applications, either by perfusing vehicle until the stimulus is removed or changing the bath solution multiple times. To gauge the required perfusion duration or number of rinses, one can perfuse or apply a colored dye and proceed to wash it out while taking aliquots of the liquid at regular intervals from the dish for several minutes. The amount of colored dye in each aliquot can be measured by a spectrophotometer and a time course of the washout can be constructed. The brain should be allowed to recover for 1–3 min after the wash out before being stimulated again. The recovery duration will depend on the time it takes for the neurons to recover a stable baseline of FRET values. If using a dipping cone objective, we recommend cleaning the dipping cone objective between recordings with deionized water and lens paper in order to remove possible residual compounds before imaging the next time-course.

Before initiating an experiment using Epac1-camps for a given GAL4/UAS combination and neuron class, it is important to confirm that the sensor produces measurable FRET changes in response to increased cAMP under the imaging parameters chosen and that these parameters support stable baseline FRET levels in vehicle controls. Stimulation of the dissected brain with 20 μM forskolin is a convenient method of confirming the appropriate conditions for successful Epac1-camps imaging. Forskolin directly activates most adenylate cyclases, thereby acutely and potently increasing cAMP production (38). If the brain is intact and the sensor is expressed at adequate levels, bath-applied 20 μM forskolin typically yields greater than 20% changes in the Epac1-camps FRET ratio (Fig. 3). For all bath-application experiments, it is also imperative to confirm that cAMP changes are not elicited in response to application of the vehicle in which a test compound is dissolved (Fig. 4). This serves as both a negative vehicle control for the compound of interest and as a control to insure that laser power, dwell time, and sampling frequency parameters support a relatively stable baseline FRET time-course (see note 8).

Compounds of interest should be dissolved solely in HL3 saline when possible. However, low final concentrations (typically around 0.1% Volume/Volume) of dimethyl sulfoxide (DMSO) are commonly used to dissolve insoluble peptides and compounds. This concentration of DMSO does not typically evoke Epac1-camps FRET changes (16). Vehicle and other negative controls must be applied in precisely the same manner as test compounds to insure that they may serve as comparable controls for movement artifacts and basal FRET changes over time. In addition to controlling for the expected pharmacological effect of a particular compound, it is equally important to control for the possible changes in pH and osmolarity that often accompany bath application of a compound. This can be accomplished through the addition of chemically similar compounds that are not predicted to have pharmacological activity.

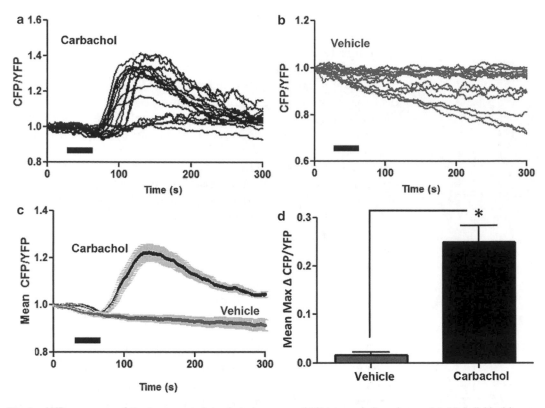

Fig. 4. cAMP responses of the large ventrolateral clock neurons (l-LNv) to a cholinergic agonist. (**a**) Individual inverse Epac1-camps FRET ratio traces recorded from single l-LNv from *w;UAS-Epac1camps/Clock-GAL4* brains before, during and after a 30 s perfusion of 10⁻⁴ M carbachol. The *black bar* indicates the time of carbachol perfusion. (**b**) Inverse FRET ratio traces for 30-s vehicle perfusions, HL3. Genotype, neuron type, and imaging parameters where identical to those for (**a**). (**c**) Averaged inverse FRET ratio traces for carbachol (*black*) and HL3 (*grey*). The *shaded* regions represent the standard error of the mean for each time-point. (**d**) Mean maximum changes in the inverse Epac1-camps FRET ratio plotted for the vehicle controls (*grey*) and carbachol treated brains (*black*). A Student's *t*-test revealed a significant difference between the two treatments ($P < 0.0001$).

3.4. Data Analysis

It is critical that all putative Epac1-camps FRET responses should be positively confirmed as true FRET changes as opposed to photobleaching artifacts (see note 9). One can check this by aligning the raw values for the CFP and YFP traces and visually checking for the antiphasic changes in CFP and YFP values that underlie true FRET responses (Fig. 3a). *True FRET responses will be characterized by increasing CFP intensities with a concomitant and simultaneous decrease in YFP values or vice versa* (Fig. 3a). Only simultaneous and opposing changes in CFP and YFP intensities signify a bonafide FRET response. Changes in the FRET ratio caused by unequal but unidirectional changes of CFP and YFP emission indicate an instance of false FRET change. In practice, this is usually caused by the uneven photobleaching of YPF and CFP, with the more rapid loss of YFP emission causing a reduction in YFP/CFP FRET values (18). Such FRET ratio drift may be unavoidable at times, but such changes should also be apparent in matched negative controls and should therefore not be mistaken for a significant cAMP response (see note 10).

Due to overlap in the emission spectra of CFP and YFP, a considerable amount of CFP emission will "spillover" into the YFP channel during FRET imaging. This spillover does not preclude the measurement of Epac1-camps FRET changes or measurement of relative cAMP changes (14). Nevertheless, many investigators use the so-called "corrected FRET method" to correct for this spillover (24), which often serves to increase the magnitude of the FRET response (Fig. 3d). For most imaging systems, the only significant spillover consists of CFP emission spillover into the YFP channel. The spillover coefficient (i.e., the proportion of CFP emission detected by the YFP channel) is unique to every imaging system. One can measure an imaging system's CFP spillover coefficient by imaging cells that strongly express CFP but not YFP using optics for YFP/CFP FRET. Under these conditions the cells will be visible in both the CFP and YFP channels despite the absence of YFP. The ratio of the mean pixel intensity of the YFP image and the CFP image is the CFP spillover coefficient. Mean pixel intensities can be measured in real time using the confocal acquisition software or using programs such as Image J (The National Institutes of Health, http://rsbweb.nih.gov/ij/), Metamorph (Molecular Devices, Inc., Sunnyvale, CA), MatLab (Mathworks, Natick, MA), or others. The CFP spillover coefficient value typically ranges from 0.5 to 0.9 for most imaging systems (18). We do not currently know of a *UAS-CFP* line that drives sufficient CFP in fly neurons for use in spillover measurements. As an alternative approach, one can transfect cell lines with CFP plasmids and use CFP expressing cells to determine spillover. Once the spillover coefficient value has been determined, the YFP intensities at each point can be spillover corrected in the following manner:

$$YFP_{SOC} = YFP - (CFP \times SO^{CFP}),$$

where YFP_{SOC} is the spillover corrected YFP intensity, CFP and YFP are the intensities from each imaging channel at a given time point, and SO^{CFP} is the spillover coefficient.

The spillover-corrected FRET ratio is then determined for each time-point as the value of YFP_{SOC}/CFP. The time-course traces of these values can be filtered and normalized in order to more easily pool data and compare the effects of different treatments (Fig. 4). We typically normalize the baseline FRET for each ROI to 1.0 by dividing all of the time-points by the FRET ratio value of the first time-point (Fig. 4a, b). For most cAMP responses, FRET ratio traces can be filtered with a moving average, whereby each ratio value is averaged with a number of values surrounding that point (Fig. 3d). The number of time-points included in the moving average, usually between five and ten data points, will depend on the nature of the data and should be determined empirically. When calculated using the standard FRET ratio of YFP_{SOC}/CFP, Epac1-camps FRET is inversely proportional to cAMP levels (39). Thus, some researchers have preferred to express Epac1-camps data in terms of the inverse FRET ratio CFP/YFP_{SOC} (Figs. 3e and 4), thereby making the relationship between Epac1-camps FRET plots and cAMP more intuitive (27). Pooled Epac1-camps data can be expressed as mean traces, wherein the mean Epac1-camps FRET value is determined for each time-point (Fig. 4c). Furthermore, mean maximum changes in the FRET ratio can then be determined for each treatment and compared statistically (Fig. 4d).

If the data are normally distributed, Student's t-tests can be used to compare the maximum FRET changes between a compound and its vehicle control. In the absence of a normal distribution, the Mann–Whitney U-test can be used. For the comparison of more than two treatments, an analysis of variance (ANOVA) with a Kruskal–Wallis test for nonparametric data can be used. Comparisons of mean traces can also be performed using a repeat measures ANOVA.

3.5. Example Experiment

Figure 4 displays a simple representative Epac1-camps experiment. In this example, we have tested whether the large ventrolateral neurons (l-LNv) of *Drosophila*'s circadian clock neuron network, display cAMP responses in response to bath-applied carbachol, a cholinergic receptor agonist. The adult l-LNv are known to express nicotinic acetylcholine receptors (40). We therefore asked if bath-applied carbachol, a potent cholinergic agonist, would have measureable effects on cAMP levels in the adult l-LNv neurons. Thirty-second perfusions of carbachol caused a consistent increase in the Epac1-camps inverse FRET ratio for the majority of neurons tested (Fig. 4a), consistent with cAMP increases. In contrast, 30-s vehicle perfusions were associated with stable inverse FRET values or a linear decrease in inverse FRET over time (Fig. 4b). Mean plots for carbachol and vehicle treated controls are shown in Fig. 4c and clearly indicate an effect of carbachol on the cAMP levels in l-LNv neurons relative to controls. In Fig. 4d the effects of cabachol and vehicle treatments are expressed as maximum percent change in the inverse FRET ratio following bath-application. We conclude from these data that bath-applied carbachol results in increased cAMP levels in the l-LNv.

3.6. Time Required

Once flies expressing the sensor are obtained (see note 11), productive recording sessions can be performed over the course of a few hours. The dissection, mounting, and recovery of each brain can take less than 15 min, while recording intervals can vary from 3 to 30 min depending on the nature of the experiment. For many experiments, one specimen can be prepared while another is being imaged. Given the variable nature of physiological experiments, we feel that seven to ten replicates should be taken for each treatment condition at minimum. Thus, for a given neuron type, testing for cAMP changes in response to one compound and its vehicle control can take less than 6 h, even for relatively long time-course experiments. In our experience, the subsequent processing and analysis of time-course data typically require an additional day's work, though such analysis can be greatly streamlined with experience and the application of appropriate software.

4. Notes

1. It is often convenient to create fly lines that stably express GAL4-driven UAS-Epac1-camps expression. Such lines obviate the need to conduct new crosses for every experiment. Any sensor that binds critical signaling molecules has the potential to interfere with cellular function, and there are unpublished reports that fly lines that widely express genetically encoded sensors can develop suppressors of sensor function. Such suppressors would be expected to interfere with Epac1-camps binding of cAMP, and thereby decrease the sensitivity of the sensor to cAMP fluctuations. Therefore, stable lines of flies expressing the sensor should be checked for sensitivity to forskolin and remade if sensitivity is determined to have diminished.

2. The brain can be most securely mounted to the bottom of the dish when all the large tracheal sacs have been removed from the exterior surface of the brain. Furthermore, the brain will adhere most securely to the dish if it is stuck down in its proper orientation on its first or second encounter with the bottom of the dish. Repeated adherence and detachment of the brain will reduce the stability of the mount. Allowing the brain to settle for 10–20 min will increase the chances of a strong adherence during live-imaging experiments. If brains fail to consistently stick securely to the bottom of the culture dish, a poly-lysine coated cover slip can be placed in the dish as an adherent surface upon which to mount the brain. The use of a coated coverslip will likely be necessary to securely mount brains if they have been treated with papain or collagenase.

3. For experiments involving a living brain, some movement of the preparation will be inevitable. The challenge is to minimize this movement so that the measurement of CFP and YFP intensities are not unduly effected by cAMP-independent events. Movement of the brain can be minimized using several approaches. The first and most critical means of minimizing

movement is the strong adherence of the brain to the bottom of the culture dish (see note 2). If this does not work, other physical devices can be constructed to immobilize brains for live-imaging experiments (41). Filtering the FRET ratios with moving averages will minimize the effects of the unavoidable subtle movements that accompany most live imaging experiments. Lastly, the method of bath-application should be optimized to result in minimal detectable movement of the preparation. For perfusion systems this will largely be a question of flow rate and smooth transitions between reservoirs.

4. There are times when GAL4 elements may drive *UAS-Epac1-camps* expression weakly, making the measurement of FRET changes difficult. In instances when there are no alternative, more strongly expressed GAL4 lines, there are several options available to the researcher to increase Epac1-Camps signals. In some cases using older flies (2–3 weeks old) can yield neurons with higher Epac1-camps expression, though these brains will likely be less robust than younger brains. In cases when the GAL4 expression pattern is relatively sparse, the diameter of confocal aperture can be increased to increase the thickness of the optical section, thereby allowing more light to be collected. This approach has the advantage of not requiring increased light stimulation. In this case the increased signal comes at the cost of Z-resolution, which might be a problem if the GAL4 element drives *UAS-Epac1-camps* in nontarget neurons residing just above or below the image's optical section. Laser powers can be increased as a means of increasing Epac1-camps brightness. But this can often result in marked false FRET changes due to an unequal bleaching of YFP and CFP. Increasing the scanning laser's dwell time (i.e., by slowing down the rate of laser scanning) is another way to increase the brightness of Epac1-camps. This approach might require a lower frequency of scans, which would decrease the temporal resolution of time-course experiments and may also result in increased photobleaching. Regardless of the adjustment made for weak sensor expression, researchers must be sure to use the same imaging parameters for both experimental and control time-courses. In general, one can maximize the apparent brightness of a neuron class's Epac1-camps expression by mounting the dissected brain in an optimal orientation. For example, if the neuron class of interest is situated just below the anterior surface of the brain, it is best to mount the brain with the anterior surface situated toward the microscope objective.

5. At some laser powers, scanning neurons too frequently can result in rapid and uneven photobleaching of CFP and YFP. For many types of cAMP responses there is no need for extremely rapid scanning. For example, scanning frequencies

of one frame every 5 s or slower were sufficient to detect the G-protein-coupled receptor signaling in the explanted fly brain (16). When high frequency scanning is necessary, care should be taken to minimize laser power to the greatest possible extent.

6. The lack of a measurable response to a given transmitter does not necessarily mean that the neuron of interest does not express receptors to the transmitter in question. There are several alternative explanations for the lack of a response. For example, the bath-applied transmitter may fail to reach the neuron of interest or the neuronal response, though biologically significant, may be below the limits of detection by the sensor. Treatment with enzymes such as papain or collagenase can be used to make the brain more permeable to bath-applied compound, but at the cost of some level of additional damage to the dissected brain. Alternatively, some compounds may be degraded or removed from extracellular spaces by endogenous mechanisms, thereby decreasing the effective concentration of the bath-applied compound. Employing high concentrations, blocking endogenous degradation/reuptake mechanisms, or using agonists not recognized by these mechanisms are all potential ways of overcoming this issue.

7. Fluctuations in the Epac1-camps FRET ratio *before* the stimulus is introduced are likely due to cAMP changes in response to the trauma of dissection and transfer to HL3 saline. To prevent this occurrence, always allow the mounted brain to recover for at least 10 min in HL3 before initiating a recording sequence.

8. Improper balance of ionic concentrations in saline can adversely affect the stability of the explanted brain and its neuronal activity. Saline should be made fresh or filtered and stored at $-20°C$ to maintain the quality of the solution. Improperly made or spoiled HL3 saline can result in poor cAMP responses and/or inconsistent recordings.

9. For some neuronal types and *GAL4/UAS-Epac1-camps* combinations, a baseline drift in FRET ratios will be unavoidable (e.g., Fig. 4b). In these cases negative controls will show significant changes in FRET over time, though these will likely be false FRET changes caused by unequal loss of YFP and CFP emission. In these cases we suggest displaying figures containing the individual plots of a given experiment. In cases where there are significant effects of a compound, the difference between compound-treated and vehicle-treated neurons will be apparent when the individual plots are compared (Fig. 4a, b). Alternatively, a mean plot of the vehicle control can be determined and the values for this plot shows over time can be subtracted from each compound-treated plot. This is a convenient means of removing FRET drift from experimental data.

10. There will be instances when rapid false FRET changes caused by a precipitous and uneven photobleaching of YFP and CFP can be easily mistaken for a significant change in Epac1-camps FRET. It is therefore extremely important that the raw CFP and YFP intensities are routinely inspected to confirm the simultaneous and opposite changes in CFP and YFP intensity that underlie bona-fide FRET responses (Fig. 3a).

11. Due to the fact that the majority of GAL4 lines were created through the random integration of P-elements into the fly genome, there are times when the very presence of the GAL4 will be accompanied by a mutation that decreases the general health of the fly or perhaps even interferes with the very neuronal circuitry one is trying to examine. Researchers should be mindful of the potential for such P-element effects.

Acknowledgments

We thank Paul Taghert and Nick Glossop for fly stocks and Ann Marie Macara and Charles Williams for comments on the manuscript. This work was supported by an NIH, NINDS grant (R00NS062953) to O.T.S.

References

1. Weiner J (1999) Time, love, memory. Vintage Books, New York

2. Simpson JH, Stephen FG (2009) Genetic dissection of neural circuits and behavior. In: Stephen F. Goodwin (ed) Advances in genetics. Academic Press (Elsevier), San Diego, CA, vol. 65, pp 79–143

3. Guerrero G, Isacoff EY (2001) Genetically encoded optical sensors of neuronal activity and cellular function. Curr Opin Neurobiol 11:601–607

4. Fiala A, Spall T, Diegelmann S, Eisermann B, Sachse S, Devaud JM, Buchner E, Galizia CG (2002) Genetically expressed Cameleon in *Drosophila melanogaster* is used to visualize olfactory information in projection neurons. Curr Biol 12:1877–1884

5. Tian L, Hires SA, Mao T, Huber D, Chiappe ME, Chalasani SH, Petreanu L, Akerboom J, McKinney SA, Schreiter ER, Bargmann CI, Jayaraman V, Svoboda K, Looger LL (2009) Imaging neural activity in worms, flies and mice with improved GCaMP calcium indicators. Nat Methods 6:875–881

6. Beavo JA, Brunton LL (2002) Cyclic nucleotide research – still expanding after half a century. Nat Rev Mol Cell Biol 3:710–718

7. Spaulding SW (1993) The ways in which hormones change cyclic adenosine 3′,5′,-monophosphate-dependent protein kinase subunits, and how such changes affect cell behavior. Endocr Rev 14:632–650

8. Borland G, Smith BO, Yarwood SJ (2009) EPAC proteins transduce diverse cellular actions of cAMP. Br J Pharmacol 158:70–86

9. Yin JCP, Tully T (1996) CREB and the formation of long-term memory. Curr Opin Neurobiol 6:264–268

10. Schild D, Restrepo D (1998) Transduction mechanisms in vertebrate olfactory receptor cells. Physiol Rev 78:429–466

11. Nakagawa T, Vosshall LB (2009) Controversy and consensus: noncanonical signaling mechanisms in the insect olfactory system. Curr Opin Neurobiol 19:284–292

12. Levine JD, Casey CI, Kalderon DD, Jackson FR (1994) Altered circadian pacemaker functions and cyclic AMP rhythms in the drosophila learning mutant dunce. Neuron 13:967–974

13. O'Neill JS, Maywood ES, Chesham JE, Takahashi JS, Hastings MH (2008) cAMP-dependent signaling as a core component of

the mammalian circadian pacemaker. Science 320:949–953

14. Vincent P, Gervasi N, Zhang J (2008) Real-time monitoring of cyclic nucleotide signaling in neurons using genetically encoded FRET probes. Brain Cell Biol 36:3–17

15. Lissandron V, Rossetto MG, Erbguth K, Fiala A, Daga A, Zaccolo M (2007) Transgenic fruit-flies expressing a FRET-based sensor for in vivo imaging of cAMP dynamics. Cell Signal 19:2296–2303

16. Shafer OT, Kim DJ, Dunbar-Yaffe R, Nikolaev VO, Lohse MJ, Taghert PH (2008) Widespread receptivity to neuropeptide PDF throughout the neuronal circadian clock network of Drosophila revealed by real-time cyclic AMP imaging. Neuron 58:223–237

17. Gervasi N, Tchénio P, Preat T (2010) PKA dynamics in a Drosophila learning center: coincidence detection by Rutabaga adenylyl cyclase and spatial regulation by Dunce phosphodiesterase. Neuron 65:516–529

18. Börner S, Schwede F, Schlipp A, Berisha F, Calebiro D, Lohse MJ, Nikolaev VO (2011) FRET measurements of intracellular cAMP concentrations and cAMP analog permeability in intact cells. Nat Protoc 6:427–438

19. Nikolaev VO, Bünemann M, Hein L, Hannawacker A, Lohse MJ (2004) Novel single chain cAMP sensors for receptor-induced signal propagation. J Biol Chem 279:37215–37218

20. DiPilato LM, Cheng X, Zhang J (2004) Fluorescent indicators of cAMP and Epac activation reveal differential dynamics of cAMP signaling within discrete subcellular compartments. Proc Natl Acad Sci USA 101:16513–16518

21. Ponsioen B, Zhao J, Riedl J, Zwartkruis F, van der Krogt G, Zaccolo M, Moolenaar WH, Bos JL, Jalink K (2004) Detecting cAMP-induced Epac activation by fluorescence resonance energy transfer: Epac as a novel cAMP indicator. EMBO Rep 5:1176–1180

22. Gloerich M, Bos JL (2010) Epac: defining a new mechanism for cAMP action. Annu Rev Pharmacol Toxicol 50:355–375

23. Willoughby D, Cooper DMF (2008) Live-cell imaging of cAMP dynamics. Nat Methods 5:29–36

24. Xia Z, Liu Y (2001) Reliable and global measurement of fluorescence resonance energy transfer using fluorescence microscopes. Biophys J 81:2395–2402

25. Lissandron V, Terrin A, Collini M, D'Alfonso L, Chirico G, Pantano S, Zaccolo M (2005) Improvement of a FRET-based indicator for cAMP by linker design and stabilization of

donor-acceptor interaction. J Mol Biol 354:546–555

26. Kula-Eversole E, Nagoshi E, Shang Y, Rodriguez J, Allada R, Rosbash M (2010) Surprising gene expression patterns within and between PDF-containing circadian neurons in Drosophila. Proc Natl Acad Sci 107:13497–13502

27. Tomchik SM, Davis RL (2009) Dynamics of learning-related cAMP signaling and stimulus integration in the Drosophila olfactory pathway. Neuron 64:510–521

28. Wu MN, Ho K, Crocker A, Yue Z, Koh K, Sehgal A (2009) The effects of caffeine on sleep in Drosophila require pka activity, but not the adenosine receptor. J Neurosci 29:11029–11037

29. Crocker A, Shahidullah M, Levitan IB, Sehgal A (2010) Identification of a neural circuit that underlies the effects of octopamine on sleep:wake behavior. Neuron 65:670–681

30. Shakiryanova D, Levitan ES (2008) Prolonged presynaptic posttetanic cyclic GMP signaling in Drosophila motoneurons. Proc Natl Acad Sci 105:13610–13613

31. Ashburner M, Roote J (2000) Laboratory culture of Drosophila. In: Sullivan W, Ashburner M, Hawley SR (eds) Drosophila protocols. Cold Spring Harbor Laboratory Press, Cold Spring Harbor, NY

32. Greenspan RJ (2004) Fly pushing, the theory and practice of Drosophila genetics, 2nd edn. Cold Spring Harbor Press, Cold Spring Harbor, New York

33. Jan LY, Jan YN (1976) Properties of the larval neuromuscular junction in Drosophila melanogaster. J Physiol 262:189–214

34. Stewart BA, Atwood HL, Renger JJ, Wang J, Wu CF (1994) Improved stability of Drosophila larval neuromuscular preparations in haemolymph-like physiological solutions. J Comp Physiol A Neuroethol Sens Neural Behav Physiol 175:179–191

35. Brand A, Perrimon N (1993) Targeted gene expression as a means of altering cell fates and generating dominant phenotypes. Development 118:401–415

36. Gummadova JO, Coutts GA, Glossop NRJ (2009) Analysis of the Drosophila clock promoter reveals heterogeneity in expression between subgroups of central oscillator cells and identifies a novel enhancer region. J Biol Rhythms 24:353–367

37. Wu JS, Luo L (2006) A protocol for dissecting Drosophila melanogaster brains for live imaging or immunostaining. Nat Protoc 1:2110–2115

38. De Souza NJ, Dohadwalla AN, Reden Ü (1983) Forskolin: a labdane diterpenoid with antihypertensive, positive inotropic, platelet aggregation inhibitory, and adenylate cyclase activating properties. Med Res Rev 3:201–219

39. Nikolaev VO, Lohse MJ (2006) Monitoring of cAMP synthesis and degradation in living cells. Physiology 21:86–92

40. McCarthy EV, Wu Y, deCarvalho T, Brandt C, Cao G, Nitabach MN (2011) Synchronized bilateral synaptic inputs to *Drosophila melanogaster* neuropeptidergic rest/arousal neurons. J Neurosci 31:8181–8193

41. Gu H, O'Dowd DK (2006) Cholinergic synaptic transmission in adult Drosophila Kenyon cells in situ. J Neurosci 26:265–272

INDEX

Jean-René Martin (ed.), *Genetically Encoded Functional Indicators*, Neuromethods, vol. 72,
DOI 10.1007/978-1-62703-014-4, © Springer Science+Business Media, LLC 2012

Printed by Publishers' Graphics LLC
BT20121002.12.29.76